How Does One
Cut a Triangle?

Second Edition

Alexander Soifer

How Does One
Cut a Triangle?

Second Edition

Forewords by

Philip L. Engel
Paul Erdős
Branko Grünbaum
Peter D. Johnson, Jr.
and Cecil Rousseau

 Springer

Alexander Soifer
College of Letters, Arts and Sciences
University of Colorado at Colorado Springs
1420 Austin Bluffs Parkway
Colorado Springs, CO 80918, USA
asoifer@uccs.edu

ISBN 978-0-387-74650-0   e-ISBN 978-0-387-74652-4
DOI 10.1007/978-0-387-74652-4
Springer Dordrecht Heidelberg London New York

Library of Congress Control Number: 2009933090

Mathematics Subject Classification (2000): 51-xx, 15-xx, 15A06, 00A07, 00A08

Cover illustration: The photographs on the front cover depict, from the upper left, clockwise Paul Erdős, John H. Conway, Matthew Kahle, and Dmytro Karabash
Cover design: Mary Burgess

Printed on acid-free paper

Springer is part of Springer Science+Business Media (www.springer.com)

To the Great Three:
   Paul Erdős,
      Isaac Yaglom,
         and my mother, Rebecca Soifer,
with eternal debt and love

Frontispiece reproduces the front cover of the original edition. It was designed by my late father Yuri Soifer, who was a great artist. Will Robinson, who produced a documentary about him for the Colorado Springs affiliate of ABC, called him "an artist of the heart." For his first American one-man show at the University of Colorado in June–July 1981, Yuri sketched his autobiography:

I was born in 1907 in the little village Strizhevka in the Ukraine. From the age of three, I was taught at the Cheder (elementary school by a synagogue), and since that time I have been painting. At the age of ten, I entered Feinstein's Jewish High School in the city of Vinniza. The art teacher, Abram Markovich Cherkassky, a graduate of the Academy of Fine Arts at St. Petersburg, looked at my book of sketches of praying Jews, and consequently taught me for six years, until his departure for Kiev. Cherkassky was my first and most important teacher. He not only critiqued my work and explained various techniques, but used to sit down in my place and correct mistakes in my work until it was nearly unrecognizable. I couldn't then touch my work and continue—this was unforgettable.

In 1924, when I was 17, my relative, the American biologist, who later won the Nobel Prize in 1952, Selman A. Waksman, offered to take me to the United States to study and become an artist, and to introduce me to Chagall, but my mother did not allow this, and I went to Odessa to study at the Odessa Institute for the Fine Arts, in the studio of Professor Mueller. Upon graduation in 1930, I worked at the Odessa State Jewish Theater, and a year later became the chief set and costume designer. In 1934 I came to Moscow to design plays for Birobidzhan Jewish Theater under the supervision of the great Michoels. I worked for the Jewish newspaper Der Emes, *the Moscow Film Studio, Theater of Lenin's Komsomol, a permanent National Agricultural Exhibition. Upon finishing 1941–1945 service in the World War II, I worked for the National Exhibition in Moscow VDNH.*

All my life I have always worked in painting and graphics. Besides portraits and landscapes in oil, watercolor, gouache, and marker (and also acrylic upon the arrival in the USA), I was always inspired (perhaps, obsessed) by the images and ideas of the Russian Civil War, World War II, biblical stories, and the little Jewish village that I came from.

The rest of my biography is in my works!

Front cover of the first edition, 1990, by Yuri Soifer.

Forewords for the Second Edition

There are very few books in the English language that can be compared to the one you are holding in your hands. Here are just a few of its attributes—taken singly, some are shared with others—but as a whole it is rather unique.

It is serious but very readable.

It poses attractive problems that are easy to understand and do not require years of study, but are far from trivial.

It challenges readers to approach mathematics as active participants, and invest time in original thinking rather than in applying ready-made tools.

Its problems have led to several published research papers, and are likely to continue doing so.

The author's genuine interest in promoting geometric and combinatorial thinking comes through in every page.

For all these reasons, I hope and expect that this new and expanded version will attract as wide and animated following as did the original edition almost two decades ago.

Branko Grünbaum
Professor of Mathematics
University of Washington
June 2008, Seattle, Washington

What can I add to the astute praise of Alexander Soifer's *How Does One Cut a Triangle?* from Philip Engel, Cecil Rousseau, Branko Grünbaum, and the late Paul Erdős? Perhaps I can provide a bit of amateur socio-historical perspective for those not familiar with Professor Soifer's extraordinary contributions.

For reasons I do not understand, mathematics in parts of Eastern Europe has developed quite a different style and flavor from mathematics in the West over the past 150 years. In Hungary, Poland, and Russia there have long been mathematical prize examinations for teenagers; in cafés in university towns mathematics students have long volleyed problems and proofs about, and beautiful problems have been as prized as beautiful proofs. Am I a romantic exaggerator? Very possibly, for I have never set foot in Hungary, Poland, or Russia, and make no claim to know what I'm talking about. But anyone would suspect from stories about Banach and Erdős and their cohorts, and the striking achievements of Eastern European mathematicians, that something quite amazing has been going on in that part of the mathematical world—perhaps not the candy box drawing I posit above, but something like it, only richer.

Professor Soifer gives us a glimpse of how it was by mixing anecdote and reminiscence with his mathematics, and that glimpse is fascinating, but his contribution is greater than that. His (partly unconscious) mission is to bring his own particular but also Eastern

European verve and style in mathematics to America, to develop the sense of beauty in problems and results, questions and answers, the glory of sheer wit in mathematical conversation, that existed and probably still exists in Moscow, Budapest, and Warsaw.

How Does One Cut a Triangle? is a work of art, and rarely, perhaps never, does one find the talents of an artist better suited to his intention than we find in Alexander Soifer and this book. On its own terms, it succeeds brilliantly. But what about on your terms? I can signal to you the presence of art, but I am not foolish enough to attempt to instruct you on how to experience it. My only advice is: sharpen your pencils and stay alert.

Peter Johnson
Professor of Mathematics
Auburn University
June 11, 2008, Auburn, Alabama

Forewords for the First Edition

How does one describe a mathematician? I know no better way than to read this delightful book. For Alexander Soifer has not just written another book about mathematics. He has opened up his heart— and in the process, his great love for mathematics, his creative ability, his respect for and skill in teaching, his joy in living all shine through. I invite you to come meet a dear friend and share his exhilaration and insight.

Philip L. Engel
Chairman
MATHCOUNTS Foundation
January 1990, Chicago, Illinois

This delightful book considers and solves many problems in dividing triangles into n congruent pieces and also into similar pieces, as well as many extremal problems about placing points in convex figures. The book is primarily meant for clever high school students and college students interested in geometry, but even mature mathematicians will find a lot of new material in it. As far as I know, the problems and results on $\mathbf{S}(F)$ (Chapter 8) are new.

Many related unsolved problems are stated in the book and the readers are encouraged to try to solve them and also to create their own problems. High school geometry and a little algebra as well as a certain amount of mathematical maturity and certainly a love for the subject are all that is needed to read the book.

I very warmly recommend the book and hope the readers will have pleasure in thinking about the unsolved problems and will find new ones.

Paul Erdős
Member of the Hungarian Academy of Sciences
Honorary Member of the National
Academy of Sciences of the USA
January 1990, Stony Brook, New York

Many people find mathematics attractive because it presents to the mind the same challenge as other activities—such as sports—present to the body. The challenge is to do the best we can with what we have.

In mathematics, and especially in geometry, there is an abundance of topics that are accessible without much previous knowledge. They present the exploring mind with opportunities to rise to that challenge, and to experience the joy of discovery. *How Does One Cut a Triangle?* is an excellent guide to this aspect of mathematics, apt to bring pleasure to anybody willing to devote a few hours to follow its adventures among solved and unsolved questions.

Branko Grünbaum
Professor of Mathematics
University of Washington
January 1990, Seattle, Washington

There is a view, held by many, that mathematics books are dull. This view is not without support. It is reinforced by numerous examples at all levels, from elementary texts with page after page of mind-numbing drill to advanced books written in a relentless Theorem–Proof style.

How Does One Cut a Triangle? is an entirely different matter. It reads like an adventure story. In fact, it is an adventure story, complete with interesting characters, moments of exhilaration, examples of serendipity, and unanswered questions. It conveys the spirit of mathematical discovery and it celebrates the event as have mathematicians throughout history.

It is said that when he discovered the famous theorem which bears his name, Pythagoras offered a hundred oxen to the Muses in thanks for the inspiration. Alexander Soifer was more restrained. When he had finished writing down the solution of Grand Problem I, he took a walk around Quail Lake to savor the moment and relax the mind before entertaining new questions which are prompted by the newly found answer. So it is with mathematics.

Cecil Rousseau
Professor of Mathematics
Memphis State University
January 1990, Memphis, Tennessee

Contents

Forewords for the Second Edition

Foreword by Branko Grünbaum ix

Foreword by Peter D. Johnson, Jr. x

Forewords for the First Edition

Foreword by Philip L. Engel xiii

Foreword by Paul Erdős xiv

Foreword by Branko Grünbaum xv

Foreword by Cecil Rousseau xvi

Preface to the Second Edition xxi

Preface to the First Edition xxvii

Part I The Original Book

1 **A Pool Table, Irrational Numbers, and Integral Independence** 3

   1.1 A Pool Table Problem 3

   1.2 Numbers Rational and Irrational 7

   1.3 Integral Independence 8

2 **How Does One Cut a Triangle? I** 15

3   Excursions in Algebra 25
    3.1   A Good Try 25
    3.2   Excursion in Linear Algebra 27
    3.3   Excursion in Algebraic Equations 34

4   How Does One Cut a Triangle? II 37

5   Excursion in Trigonometry 41

6   Is There Anything Beyond the Solution? 47

7   Pursuit of the Best Result............................ 51
    7.1   Birth of a Problem 51
    7.2   The Optimal Result 55
    7.3   Second Solution to the Five-Point Problem 60

8   Convex Figures and the Function $S(F)$ 65
    8.1   Creating a Function 65
    8.2   Study of Convex Figures:
          The Upper Bound of $S(F)$ 70
    8.3   Study of Convex Figures:
          A Lower Bound of $S(F)$ 77
    8.4   A Very Brief Introduction to
          Affine Geometry 79
    8.5   Study of Convex Figures:
          The Best Lower Bound of $S(F)$ 86
    8.6   A One-Hundred-Dollar Problem 96
    8.7   Triangle in Ellipse............................... 99

9   Paul Erdős: Our Joint Problems 107
    9.1   PGOM Erdős 107
    9.2   Problems.. 109
    9.3   Solutions... 110

10 Convex Figures and Erdős' Function $S_\alpha(F)$ 121

Part II   Developments of the Subsequent 20 Years

11   An Alternative Proof of Grand Problem II 127

12   Miklós Laczkovich on Cutting Triangles 129

13   Matthew Kahle on the Five-Point Problem 137

14   Soifer's One-Hundred-Dollar Problem and
     Mitya Karabash 143

15   Coffee Hour and the Conway–Soifer Cover-Up 147

16   Farewell to the Reader............................... 157

Appendix A   Cutting a Triangle into Congruent Triangles . 161

Appendix B   The Five-Point Problem 163

References 167

Notation 171

Index 173

Preface to the Second Edition

The development of science and of creative activities of the spirit in general requires still another kind of freedom, which may be characterized as inward freedom. It is this freedom of the spirit which consists in the independence of thought from the restrictions of authoritarian and social prejudices as well as from unphilosophical routinizing and habit in general.
—*Albert Einstein,* On Freedom, *1940* [Ei]

A course of instruction will be the more successful the more its individual phases assume the character of experience.
—*Hugo von Hofmannsthal,* Buch der Freunde [H]

Out of all my books released prior to *Mathematical Coloring Book*, this one is my favorite. She lived with me for 20 years, since the idea was born in 1970, until her leaving me in 1990. Let us trace here highlights of the book's life after she went to live on her own. In this new edition, I am adding five chapters to the existing ten. In order to preserve the smooth flow of the original book, I add these new chapters as Part II. Part I remains unchanged, except, of course, for the correction of a number of typos.

Shortly after the book's release, the Hungarian mathematician Miklós Laczkovich informed me that the book inspired him to publish a paper on cutting triangles. For this edition, Miklós donated a note with an important new result—you will find it in Chapter 12.

At the eleventh hour, when this new edition was already in production, the winner of the 1990 and 1991 Colorado Mathematical Olympiads, and now post-doctoral fellow at Stanford University

Matthew Kahle submitted to *Geombinatorics* an essay [Kah] dedicated to my round-numbered birthday. In it, Matthew improves the result of the Five-Point Problem 7.2.1, poses further problems and formulates new conjectures. I dedicate Chapter 13 to his new work.

In 2005, the brilliant young freshman from Columbia University Mitya Karabash came to the University of Colorado at Colorado Springs, where I supervised his summer research. We looked at problems of tiling and covering, the chromatic number of the plane, and problems posed by Paul Erdős and me in this book.

To my delight, on June 4, 2007, Mitya delivered to me the most important result related to this book in the two decades that have passed since the release of the first edition. What exactly did Mitya prove? He did not solve the One-Hundred-Dollar Problem (8.6.6) that asks to classify all convex figures F for which $\mathbf{S}(F) = 6$. But he settled conjecture 8.6.7 in the negative by constructing a figure F such that $\mathbf{S}(F) = 6$, yet F cannot be mapped by an affine transformation into a however narrow frame formed by two concentric regular pentagons. The structure of the figures F with $\mathbf{S}(F) = 6$ appears to be even harder than I imagined twenty years ago. Now we do not even have a working conjecture for the One-Hundred-Dollar Problem! Mitya's results have appeared as a sequence of two papers [Ka1] and [Ka2] in the quarterly *Geombinatorics*. Chapter 14 of this book is dedicated to these results.

In 2004 I posed two easy-to-formulate, hard-to-solve problems: Cover-Up and Cover-Up Squared. In Chapter 15 you will meet these problems and see how the progress was obtained during coffee hours at Princeton University by John H. Conway and me, and then by Mitya Karabash and me, and finally by Fan Chung and Ronald L. Graham. Plenty is left open for you to join in!

The book received kind notes from many journals and Paul Erdős, Leroy M. Kelly, Murray K. Klamkin, and others. The Russian geometer Vladimir G. Boltyanski in *SIAM Review* has described my goals perhaps better than I could:

> We do not often have possibilities to look into a creative workshop of a mathematician. Usually mathematicians give an account of their results in a ground-out and logically

irreproachable form. But their creative pains, methods of investigation and means of obtaining results remain vague, especially for other mathematicians. And therefore every possibility to observe creative methods employed by a mathematician is interesting and useful. In particular, it is important for a beginner, that is, for a schoolboy or a schoolgirl who is interested in mathematics. First of all, it is important because pupils (and even first-year students in mathematical or technical colleges), as a rule, do not imagine what modern mathematics is, its scientific problems and its methods of investigation. School mathematics is far from authentic modern science with respect to its contents and methods.

... The beginner, who is interested in the book, not only comprehends a situation in a creative mathematical studio, not only is exposed to good mathematical taste, but also acquires elements of modern mathematical culture. And (not less important) the reader imagines the role and place of intuition and analogy in mathematical investigation; he or she fancies the meaning of generalization in modern mathematics and surprising connections between different parts of this science (that are, as one might think, far from each other) that unite them...

This makes the book alive, fresh, and easily readable. Alexander Soifer has produced a good gift for the young lover of mathematics. And not only for youngsters; the book should be interesting even to professional mathematicians.

Leroy M. Kelly observed the book's spirit in *The Mathematical Reviews:*

It is impossible to convey the spirit of the book by merely listing the problems considered or even a number of solutions. The manner of presentation and the gentle guidance toward a solution and hence to generalizations and new problems takes this elementary treatise out of the prosaic and into the stimulating realm of mathematical creativity. Not only young talented people but dedicated secondary teachers and even a few mathematical sophisticates will find this reading both pleasant and profitable.

John Baylis wrote in *The Mathematical Gazette* (UK):

Alexander Soifer is a wonderful problem solver and inspiring teacher. His book will tell young mathematicians what mathematics should be like, and remind older ones who may be in danger of forgetting. This review has the simple aim of persuading as many people as possible to read it...

So why am I urging you to read this? Mainly because it is such a refreshing book. Professor Soifer makes the problems fascinating, the methods of attack even more fascinating, and the whole thing is enlivened by anecdotes about the history of the problems, some of their recent solvers, and the very human reactions of the author to some beautiful mathematics. Most of all, the book has charm, somehow enhanced by his slightly eccentric English, sufficient to carry the reader forward without minding being asked to do rather a lot of work. I am tempted to include several typical quotations but I'll restrain myself to two: From Chapter 8 "Here is an easy problem for your entertainment. Problem 8.1.2. Prove that for any parallelogram P, $S(P) = 5$. Now we have a new problem, therefore we are alive! And the problem is this: what are all possible values of our newly introduced function $S(F)$? Can the function $S(F)$ help us to classify geometry figures?"

And from an introduction by Cecil Rousseau:

"There is a view, held by many, that mathematics books are dull. This view is not without support. It is reinforced by numerous examples at all levels, from elementary texts with page after page of mind-numbing drill to advanced books written in a relentless Theorem–Proof style.

"How Does One Cut a Triangle? is an entirely different matter. It reads like an adventure story. In fact, it is an adventure story, complete with interesting characters, moments of exhilaration, examples of serendipity, and unanswered questions. It conveys the spirit of mathematical discovery

and it celebrates the event as have mathematicians throughout history."

And this isn't just publishers going over the top—it's all true!

I thank Colonel Dr. Robert Ewell for converting some of my designs into illustrations in the new chapters of the book.

I am so very grateful to the first readers of this new edition: Branko Grünbaum and Peter D. Johnson, Jr., for their comments and forewords.

I am deeply grateful to Ann Kostant for inviting this book's new edition into the historic Springer.

I have good news for those who may be interested in my books: [S1], [S3] and [BS] will soon appear in new expanded editions in Springer. [S5] after 18 years in the writing came out in November of 2008. I hope my new books [ES] and [S6] will appear in Springer soon.

I welcome feedback—your problems, solutions, ideas.

Alexander Soifer

Colorado Springs, Colorado

May 8, 2008

Preface to the First Edition

If I see a really nice proof, I say it comes straight from the Book. . . God has a transfinite Book, which contains all theorems and their best proofs, and if He is well intentioned toward those, He shows them the Book for a moment. And you wouldn't even have to believe in God, but you must believe that the Book exists.

—*Paul Erdős*

How Does One Cut a Square?

—*Isaac Yaglom*

In 1970, I visited my professor, Isaac Moiseevich Yaglom, in his Moscow apartment. We talked about art and mathematics, two passions we share. Professor Yaglom complimented me on two problems about cutting triangles into similar and congruent triangles that I had just created and solved. A famous geometer, he was impressed with how algebra was used to obtain genuinely geometric results. "A spirit of the time," he said and added, "Nobody would have thought of these solutions a few years ago." Prof. Yaglom offered to publish these solutions in the proceedings of our Institute so that he could refer to them and add them as an appendix to an upcoming Springer-Verlag translation of his book, *How Does One Cut a Square?* Arrogantly, I turned this generous offer down. "Nobody reads these proceedings," I said. Little did I know that Yaglom was their Editor-in-Chief! Yet, Prof. Yaglom modestly replied, "You are probably right." I promised him, "I will write my own book, *How Does One Cut a Triangle?* with the title of your book, *How Does One Cut a Square?* as the epigraph." Now, I am making good on my promise, even after twenty years that took me from a university senior in Moscow to a mathematics professor in Colorado Springs.

Three years ago I lost my mother. She gave me life. She served as a role model of loyalty, integrity, enthusiasm, and love. A year and a half ago we all lost the famous Russian geometer and author of numerous great books, Isaac Moiseevich Yaglom. I grew up on his and Martin Gardner's books. At about the same time I met a legend, Paul Erdős. Our meetings and correspondence have inspired me to write this book and other works. This is why I am dedicating the book to these three great people.

This book is a research monograph on my newly created and solved problems of combinatorial geometry. Yet, it is not an ordinary research monograph, written for a narrow group of professionals. This book might be of interest to professional geometers, but I wrote it first of all for young, talented mathematicians who are still in high school or early college, and for their teachers, and for the leisure of all professional mathematicians. In a normal monograph, the main emphasis is on results obtained. I would not have written this book if the solutions were not beautiful and valuable in their own right.

I cannot help but quote I. M. Yaglom's preface from his book, *How Does One Cut a Square?* He so precisely describes my goals and aspirations:

> Here we have a certain "fragment of mathematics" that characterizes mathematical trains of thought, techniques, and methods. The author was guided precisely by this thought: it is not the results that are of interest here so much as the reasoning leading to those results—not the "what" of the proofs that deserves attention so much as the "how."

My goal in this book is to show young, talented people what mathematics is and what mathematicians do. I cannot overemphasize the importance of this goal; our youth throughout the world too often graduate from high schools without the foggiest idea of what mathematics is. Many university and college graduates do not know either. Indeed, too often teachers do not know, so how on earth can we expect our students to find out?

Mathematics in school is reduced to a topic study, filled with "prove that A implies B by using the Pythagorean theorem." *This book is a real, live "fragment of mathematics" with analytical proofs and constructions of counterexamples; with open and open-ended problems; with mathematical intuition leading research like a light at the end of a tunnel; with the synthesis of ideas from algebra, geometry, trigonometry, linear algebra, mathematical analysis; with beauty, elegance, and surprises.*

It is a special monograph in that I had to raise my pedagogical talent to the limit in an attempt to introduce to my young colleagues ideas from linear algebra and analysis normally studied later in college, and do it without sacrificing the rigor of mathematical reasoning. I kept some proofs out, but only those that would have led us too far from the main topic of our research or that would have made this book much thicker (I, too, hate the user-unfriendly look of calculus bricks!).

I tried to show that mathematics is alive—that every solved problem gives birth to myriad unsolved ones. This book is full of open problems. Many of them carry ten to fifty dollar prizes for their first solutions. Do not be afraid to try to be the first to solve a problem. As the *South Carolina Reflector* put it, "Think like a tea bag. You don't know your strength until you get in hot water."

This book was written in a dialogue with my colleagues, young and not so young. I thank them all. I am especially grateful to those whose solutions I included in this book: Dr. Semion Slobodnik from Moscow; Royce Peng, a high school student from California; Professor Cecil Rousseau from Memphis State University; Volodia Baranovski, a high school student from Siberia; and Boria Dubrov, a high school student from Minsk, USSR. I thank Professor Branko Grünbaum from the University of Washington for contributing an exciting open problem, Problem 8.7.10 and Conjecture 8.7.11.

Paul Erdős created problems 6.6 and 6.7, co-created all the problems of Chapter 9, and generously contributed a whole chapter of open problems in letters sent upon his reading my manuscript (Chapter 10). My hat goes off to him!

I am grateful to George Berzsenyi, Phillip Engel, Paul Erdős, Martin Gardner, Branko Grünbaum, and Cecil Rousseau for being the first readers of the manuscript and providing me with most valuable feedback. I am honored that Philip Engel, Paul Erdős, Branko Grünbaum, and Cecil Rousseau have written forewords for this book.

I thank and applaud my dean, Dr. James A. Null, for fostering such a wonderful climate for creative work in the College of Letters, Arts and Sciences. And I thank my friend and secretary, Lynn Scott, for encoding my multilayered, hand-written manuscript. "It reads like a long letter," she said.

Well, here is my letter to you, my reader! Write back to me! I'm looking forward to your solutions and new problems.

Alexander Soifer
Colorado Springs, Colorado
January 1990

Part I
The Original Book

1

A Pool Table, Irrational Numbers, and Integral Independence

1.1 A Pool Table Problem

Pool Table Problem I.

A rectangular $p \times 2q$ pool table has pockets in every corner and in the middle of each $2q$-long side (see Figure 1.1). A ball is rolled from a corner pocket at a $45°$ angle with respect to the side rails.

Find necessary and sufficient conditions on the real numbers p and q for the ball to eventually get into a pocket (angle of incidence is equal to the angle of reflection).

Solution. Every time the ball hits a rail, we can reflect the pool table symmetrically with respect to that rail. Since the angle of incidence is equal to the angle of reflection, the reflected trajectory of the ball will be a straight line! (See Figures 1.2, 1.3, and 1.4.)

As a result, our problem becomes equivalent to the following:

What is the necessary and sufficient condition for a straight line drawn from the origin at a $45°$ angle with respect to the coordinate axes to pass through a point with coordinates (mp, nq) for some positive integers m, n (see Figure 1.5)?

Since the equation of our straight line is $y = x$, the required condition is $mp = nq$, or

A. Soifer, *How Does One Cut a Triangle?*, Second Edition, DOI 10.1007/978-0-387-74652-4_1,
© Alexander Soifer 2009

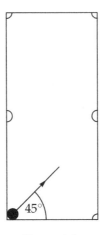

Figure 1.1

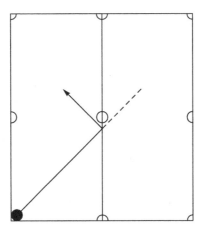

Figure 1.2

$$pm + q(-n) = 0 \qquad (*)$$

for some positive integers m, n.   □

Let us take a closer look at the equation

$$px + qy = 0.$$

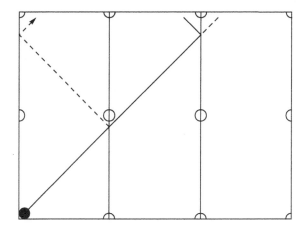

Figure 1.3

It certainly has a solution at $x = 0, y = 0$. In addition, the condition ($*$) above requires this equation to have at least one more solution in the integers, namely $x = m, y = -n$.

Such situations appear quite often in mathematics. We have a special definition for them:

If the equation

$$px + qy = 0 \qquad (1)$$

has a non-zero solution in integers, then the numbers p, q are called *integrally dependent*.

If equation (1) above has only the zero solution $x = y = 0$, then the numbers p, q are called *integrally independent*.

In Pool Table Problem I, the ball will get into a pocket of a $p \times 2q$ pool table if and only if the system of numbers p, q is integrally dependent!

Try your pool table skills on the next problem:

Pool Table Problem II.

In the setting of Pool Table Problem I, let p and q be odd integers. Prove that the ball will get into a pocket. Moreover, it will be a side pocket!

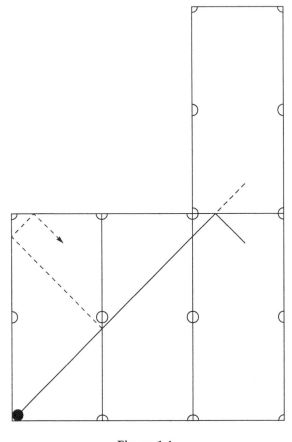

Figure 1.4

Find necessary and sufficient conditions on p, q for the ball to get into a side pocket.

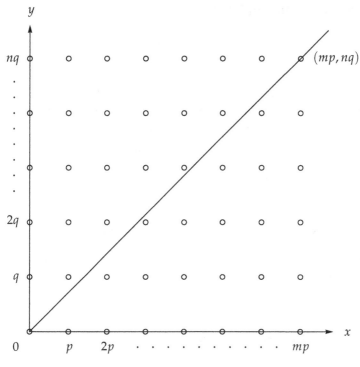

Figure 1.5

1.2 Numbers Rational and Irrational

You may recall that $\sqrt{2}$ is an irrational number, i.e., it may *not* be presented in the form $\dfrac{m}{n}$ where m and n are integers and $n \neq 0$. So,

$$\sqrt{2} \neq \frac{m}{n}$$

or

$$\sqrt{2}\,n + 1(-m) \neq 0.$$

In other words, the equation

$$\sqrt{2}\,x + 1y = 0$$

has a unique solution in integers

$$x = y = 0.$$

I.e., 1 and $\sqrt{2}$ are integrally independent!

We come again to the same *Diophantine equation* (1) (an equation in two or more variables to be solved in the integers) and to the same notion of integral independence of numbers.

Now you can prove the following statement:

Problem 1.2.1. A number i is irrational if and only if the numbers 1, i are integrally independent.

It carries, of course, the following corollary:

Problem 1.2.2. A number i is rational if and only if the numbers 1, i are integrally dependent.

1.3 Integral Independence

We are ready to generalize the definition of integral independence:

Numbers $a_1, a_2, \ldots, a_n$ are said to be *integrally independent* if the Diophantine equation

$$a_1 x_1 + a_2 x_2 + \cdots + a_n x_n = 0 \tag{2}$$

has a unique solution in integers

$$x_1 = x_2 = \cdots = x_n = 0.$$

If equation (2) above has at least one more solution, the numbers $a_1, a_2, \ldots, a_n$ are called *integrally dependent*.

Problem 1.3.1. Prove that the numbers $\sqrt{2}$, $\sqrt{3}$, $\sqrt{4}$ are integrally independent.

Solution. Let x, y, z be a solution to the Diophantine equation

$$\sqrt{2}x + \sqrt{3}y + \sqrt{4}z = 0. \tag{3}$$

If $x = 0$, then we get

$$\sqrt{3}y + 1(2z) = 0. \tag{4}$$

Since $\sqrt{3}$ is an irrational number, we know from Problem 1.2.1 that $\sqrt{3}$ and 1 are integrally independent. Thus, $y = z = 0$ is the only solution to equation (4).

Similarly, we can show that if $y = 0$, then $x = z = 0$.

Now we can assume that $x \neq 0$ and $y \neq 0$. We get from (3) that

$$\sqrt{2}x + \sqrt{3}y = -2z.$$

Squaring both sides produces

$$2x^2 + 2\sqrt{6}xy + 3y^2 = 4z^2,$$

i.e.,

$$\sqrt{6}(2xy) + 1(2x^2 + 3y^2 - 4z^2) = 0. \tag{5}$$

If you are familiar with my book *Mathematics as Problem Solving*, you may have proven that $\sqrt{6}$ is an irrational number ([S1] Problem 1.3.5). Otherwise, prove it now.

Due to the statement of Problem 1.2.1, the numbers $\sqrt{6}$ and 1 are integrally independent; thus, the equality (5) implies

$$2xy = 0$$

and

$$2x^2 + 3y^2 - 4z^2 = 0.$$

The first one requires $x = 0$ or $y = 0$, in contradiction with our assumption.

We proved that $x = y = z = 0$ is the only integral solution to equation (3); therefore, $\sqrt{2}, \sqrt{3}, \sqrt{4}$ are integrally independent. $\square$

Problem 1.3.2. Prove that three angles of a right triangle are integrally dependent.

Solution. If α, β, γ are the angles of a right triangle with $\gamma = \dfrac{\pi}{2}$, then

$$\alpha + \beta = \gamma,$$

i.e.,

$$\alpha(1) + \beta(1) + \gamma(-1) = 0.$$

The last equality shows that the equation

$$\alpha x + \beta y + \gamma z = 0$$

has a non-zero solution in integers, and thus completes the proof.　□

I am omitting solutions to the following several problems to allow you, my reader, the pleasure of solving them on your own.

Problem 1.3.3. Prove that $\sqrt{2}$, $\sqrt{3}$, and $180 - \sqrt{2} - \sqrt{3}$ are integrally independent.

Problem 1.3.4. Prove that the three angles of an isosceles triangle are integrally dependent.

Problem 1.3.5. Prove that if two numbers A, B are integrally dependent, then three numbers A, B, C are integrally dependent regardless of the value of C.

Corollary 1.3.6. If three numbers A, B, C are integrally independent, then any two of them are integrally independent as well.

Problem 1.3.7. Let the real numbers A, B, and β be connected by the following equalities:

$$A = a_{12}\beta$$
$$B = a_{22}\beta$$

where a_{12} and a_{22} are integers.
Then A, B are integrally dependent.

Problem 1.3.8. Let the real numbers A, B, C, α, and β be connected by the following equalities:

$$A = a_{11}\alpha + a_{12}\beta$$
$$B = a_{21}\alpha + a_{22}\beta \qquad (6)$$
$$C = a_{31}\alpha + a_{32}\beta$$

where all a_{ij} are integers.

Then the three numbers A, B, C are integrally dependent.

Hint: If $a_{11} \cdot a_{22} = a_{21} \cdot a_{12}$, then

$$Aa_{21} + B(-a_{11}) + C \cdot 0 = 0,$$

i.e., the equation

$$Ax + By + Cz = 0 \qquad (7)$$

has a non-zero integral solution.

If $a_{11} \cdot a_{22} \neq a_{21} \cdot a_{12}$, then you can solve the first two equalities of (6) for α and β, and then substitute these solutions for α and β into the third equality of (6). You will get a non-zero integral solution of the equation (7).

Problem 1.3.9. Prove the statement of Problem 1.3.8 if all coefficients a_{ij} are allowed to be rational numbers.

Problem 1.3.10. Let the numbers $a_1, a_2, \ldots, a_n$ be integrally independent. Then for integers $x_1, x_2, \ldots, x_n; y_1, y_2, \ldots, y_n$, the equality

$$x_1 a_1 + x_2 a_2 + \cdots + x_n a_n = y_1 a_1 + y_2 a_2 + \cdots + y_n a_n$$

is true if and only if the corresponding coefficients are equal, i.e., if

$$x_1 = y_1$$
$$x_2 = y_2$$
$$\cdots$$
$$x_n = y_n.$$

Problem 1.3.11. Little grooves of the same width are dug across a long (very long!) straight road. The distance between the centers of any two consecutive grooves is a (Figure 1.6). Prove that no matter how narrow the grooves are, a man walking along the road with a step equal to b will sooner or later step into a groove, provided that the numbers a, b are integrally independent (we assume that the man's "feet" are so small that his footprints look like dots).

What if the numbers a, b are integrally dependent?

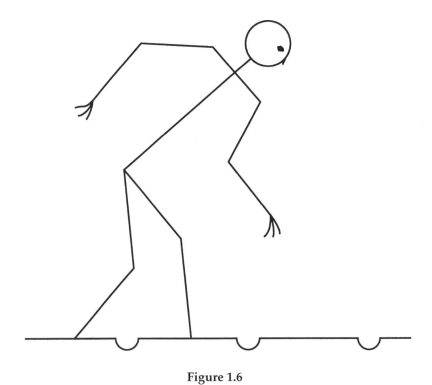

Figure 1.6

Hint: This statement is a generalization of Problem 1.4.9 from my *Mathematics as Problem Solving* [S1].

This is not an easy problem. It is not necessary to solve it in order to understand the rest of this book. I hope, though, that you, my reader, are not going to give up too quickly. Reading the solution of Problem 1.4.4 from [S1] will provide you with a necessary insight.

2

How Does One Cut a Triangle? I

You are familiar with a picture of three midlines drawn in an arbitrary triangle T (see Figure 2.1).

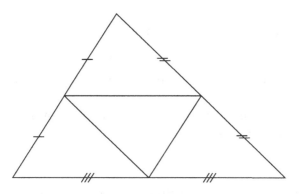

Figure 2.1

The midlines cut T into four triangles congruent to each other. You probably know that this construction can be easily generalized to all perfect squares:

> For every positive integer n, every triangle T can be cut into n^2 congruent triangles.

A. Soifer, *How Does One Cut a Triangle?*, Second Edition, DOI 10.1007/978-0-387-74652-4_2,
© Alexander Soifer 2009

All we have to do is partition each side of T into n segments of equal length (see Figure 2.2) and connect the corresponding marks of partitions by lines parallel to the sides of the triangle T (you should prove that we indeed get n^2 congruent triangles).

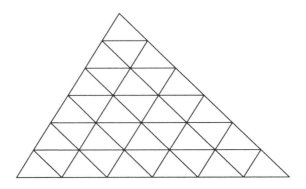

Figure 2.2

We arrive at our first major problems:

(1) Are there any other numbers n besides the perfect squares such that every triangle can be cut into n triangles congruent to each other?

(2) What would be the answer if we relax the requirement of congruency to similarity?

In other words:

Grand Problem I. Find all positive integers n such that every triangle can be cut into n triangles congruent to each other.

Grand Problem II. Find all positive integers n such that every triangle can be cut into n triangles similar to each other.

It appears that if anything, the first problem is easier than the second. The opposite, in fact, is true. We have all the tools necessary to solve Grand Problem II here. Problem I will have to wait until we build some additional instruments in the next chapter.

Let us *try* to solve Grand Problem II.

Figure 2.2 shows that every triangle can be cut into n^2 triangles similar to each other. By combining all $(n-1)^2$ triangles located above the lowest cut parallel to the base, we get Figure 2.3 from Figure 2.2.

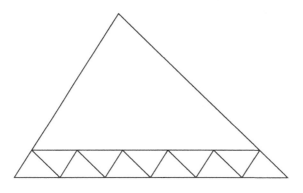

Figure 2.3

Figure 2.3 shows a partition of a triangle into $n^2 - (n-1)^2 + 1$ triangles similar to each other, but

$$n^2 - (n-1)^2 + 1 = 2n.$$

Therefore, every triangle can be cut into any *even* number of similar triangles greater or equal than 4.

What about odd numbers? Well, let us start with the partition of Figure 2.1 and cut one of the four triangles into $2n$ similar triangles, as in Figure 2.3. We will get $2n+3$ similar triangles (Figure 2.4). Not only is $2n+3$ an odd number, but every positive odd number greater than 5 can be expressed (prove it) in the form $2n+3$, where n is an integer greater than 1.

Thus far we have proven that every triangle can be cut into n triangles similar to each other for all n except the first three prime numbers

$$n = 2, 3, 5,$$

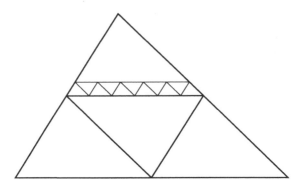

Figure 2.4

for which our constructions in Figures 2.3 and 2.4 do not work.

Will something else work?

An insightful reader may have sensed that we introduced integral independence in Chapter 1 for a reason. Indeed, *assume that a triangle T with integrally independent angles A, B, C is cut into a number of triangles with angles α, β, γ, similar to each other* (see Figure 2.5).

The partition of the triangle T, of course, induces the partitions of its angles A, B, C into the sums of angles congruent to α, β, γ (see Figure 2.5):

$$A = a_{11}\alpha + a_{12}\beta + a_{13}\gamma$$
$$B = a_{21}\alpha + a_{22}\beta + a_{23}\gamma \qquad (8)$$
$$C = a_{31}\alpha + a_{32}\beta + a_{33}\gamma,$$

where all the coefficients a_{ij} are non-negative integers.

For $j = 1, 2, 3$ we denote $\Sigma_j = a_{1j} + a_{2j} + a_{3j}$ (in other words, Σ_1, Σ_2, and Σ_3 are the sums of columns in the system of equalities (8)). Then by adding up the three equalities of (8) and remembering that $A + B + C = \pi$, we get:

$$\pi = \Sigma_1\alpha + \Sigma_2\beta + \Sigma_3\gamma. \qquad (9)$$

Let us consider two cases:

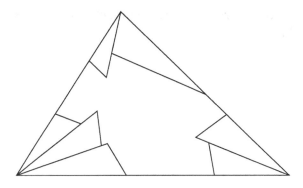

Figure 2.5

Case 1. At least one of the numbers $\Sigma_1, \Sigma_2, \Sigma_3$, say Σ_3, is equal to zero.

$\Sigma_3 = 0$ implies $a_{13} = a_{23} = a_{33} = 0$ (since all a_{ij} are non-negative), and system (8) simplifies to coincide with system (6) in Chapter 1. Due to the statement of Problem 1.3.8, (6) implies that the numbers A, B, C are integrally dependent, which contradicts our assumption.

Therefore, the following case takes place:

Case 2. Each of the numbers $\Sigma_1, \Sigma_2, \Sigma_3$ is positive. Then $\Sigma_1 \alpha + \Sigma_2 \beta + \Sigma_3 \gamma \geq \alpha + \beta + \gamma = \pi$, and due to equality (9), we conclude that

$$\Sigma_1 = \Sigma_2 = \Sigma_3 = 1.$$

This equality implies (show how!) that up to a permutation of symbols α, β, γ, system (8) simplifies to look as follows:

$$A = \alpha$$
$$B = \beta \qquad (10)$$
$$C = \gamma.$$

The equalities of (10) have two very important consequences:

(1) triangles of the partition are not only similar to each other; they are similar to the original triangle T;

(2) the angles of the triangle T were in fact not split; they were cut
     off (see Figure 2.6).

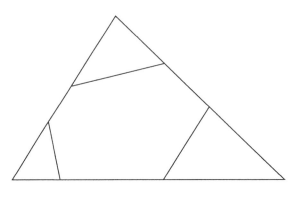

Figure 2.6

We did not solve our Grand Problem II (yet!). We did prove a
powerful result, not very important by itself but instrumental in
solving Grand Problem II and useful in other problems (such as
Grand Problem I!).

In mathematics such results are called *lemmas*. I prefer to call
them *tools*.

Tool 2.1. If a triangle T with integrally independent angles is cut
into a number of triangles T_n similar to each other, then:

(1) each triangle T_n is similar to T;

and

(2) the angles of T are not split, but rather cut off (as in Figure 2.6).

Tool 2.1 immediately proves that a triangle T with integrally independent angles cannot be cut into 2 or 3 similar triangles; indeed,
Figure 2.6 shows that we must get at least 4 pieces.

Assume that a triangle with integrally independent angles is cut
into 5 similar triangles. The middle piece of the partition in Figure

2.6 must be at most a quadrilateral, since by Tool 2.1 we already cut off the angles of T three triangles similar to T.

Thus, the middle piece is a triangle or a quadrilateral (see Figures 2.7 and 2.8).

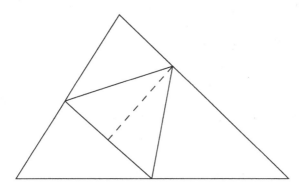

Figure 2.7

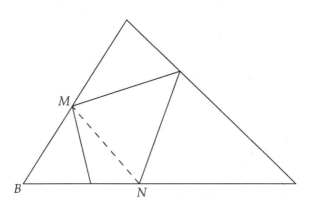

Figure 2.8

We need a little extra tool, which you can easily prove on your own:

Tool 2.2. If a triangle is cut into two triangles T_1 and T_2 similar to each other, then T_1 and T_2 are right triangles.

Now, if the middle piece were a triangle, as in Figure 2.7, then it must have been cut into two triangles T_1 and T_2 similar to each other. By Tool 2.2, T_1 and T_2 are right triangles. This means that by Tool 2.1, all five triangles of the partition, as well as the original triangle T, are right triangles. But the angles of the right triangle are integrally dependent (see Problem 1.3.2), which contradicts our assumption that the angles of T are integrally independent.

If the middle piece were a quadrilateral as in Figure 2.8, then it must have been cut into two triangles by one of its diagonals. But this diagonal (MN in Figure 2.8) cuts off the original triangle T a triangle (MNB in Figure 2.8) that is split into two triangles similar to each other. From here we travel to a contradiction by exactly the same way as in the previous case (when the middle piece was a triangle).

You, my reader, may think that Grand II is solved. In fact we "only" proved that a triangle T with integrally independent angles can not be cut into 2, 3, or 5 triangles similar to each other. But what if such a triangle T does not exist?

Cheer up: it exists! Moreover, you proved it in your home entertainment, Problem 1.3.3: the angles of measure $\sqrt{2}°$, $\sqrt{3}°$, and $(180 - \sqrt{2} - \sqrt{3})°$ are integrally independent.

Grand II is done. *Every triangle can be cut into any number n of triangles similar to each other, except the first three primes: 2, 3, and 5.*   □

I created Grand II in April 1970, when I served as one of the judges of the Soviet Union National Mathematical Olympiad. The judges liked the problem. They selected the critical part of it for the juniors (ninth graders) competition:

Can every triangle be cut into five triangles similar to each other?

Then came the meeting to approve the problems with the Chairman of the Organizing Committee, Andrej Nikolaevich Kolmogorov, one of the greatest mathematicians of the twentieth century.

Kolmogorov quietly listened to the presentation of all the problems and their solutions, and then said:

"I would only like to replace the problem about five similar triangles."

"Why, Andrej Nikolaevich?" asked the head judge for the ninth grade, Yuri Ionin.

"It is too difficult. I am not sure I would have solved it," replied Kolmogorov.

The reason was found valid, and the problem was replaced. Of course, I was not thrilled about it. And yet, in a way, I was satisfied. The great Kolmogorov thought Grand II was a difficult problem. "What would he think then about Grand I?" I exclaimed to myself.

Nineteen years later, Grand II has finally made it into the Mathematical Olympiad of the International Summer Institute at Oakdale, Long Island, New York. The competitors included some remarkable high school students from the Soviet Union, France, Switzerland, and the United States. The winner, Vania Arzhantsev, a senior from Kiev and a candidate to the Soviet team for the 1990 International Mathematical Olympiad in Beijing, was the only one to substantially advance in the problem.

His proof has an important consequence: integral independence of angles is not a necessary condition. He used a triangle with the angles $1°$, $6°$, and $173°$ as an example of a triangle that cannot be cut into five triangles similar to each other. I will share details of this solution with you in Chapter 11.

3

Excursions in Algebra

3.1 A Good Try

Let us try to solve Grand Problem I. You may recall (Figure 2.2) that every triangle can be cut into n triangles congruent to each other if n is a perfect square (i.e., $n = 1, 4, 9, 25, \dots$). The "only" question that remains is whether a partition into n congruent triangles exists *only* for perfect squares n.

Tool 2.1 worked very nicely for us in our battles with Grand II. Let us try to use it here as well. Assume that a triangle T with integrally independent angles is cut into n copies of the triangle T_1. Then by Tool 2.1, the triangle T_1 is similar to T. The area of the triangle T is n times greater than the area of T_1. If the lengths of the sides of T_1 are a, b, and c, then, of course, the lengths of the sides of T are $a\sqrt{n}$, $b\sqrt{n}$, $c\sqrt{n}$ (the ratio of linear sizes of two similar triangles is the square root of the ratio of their areas).

The partition of the triangle T into n copies of T_1 induces partitions of the sides of T into the sums of the sides of T_1 (see Figure 3.1).

We get the following system of equalities:

$$a\sqrt{n} = a_{11}a + a_{12}b + a_{13}c$$
$$b\sqrt{n} = a_{21}a + a_{22}b + a_{23}c \qquad (11)$$
$$c\sqrt{n} = a_{31}a + a_{32}b + a_{33}c$$

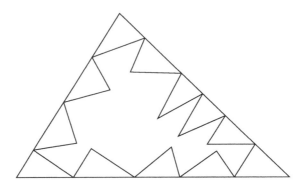

Figure 3.1

We can rewrite (11) in matrix form as follows[1]:

$$\begin{pmatrix} a \\ b \\ c \end{pmatrix} \cdot \sqrt{n} = \begin{pmatrix} a_{11} & a_{12} & a_{13} \\ a_{21} & a_{22} & a_{23} \\ a_{31} & a_{32} & a_{33} \end{pmatrix} \cdot \begin{pmatrix} a \\ b \\ c \end{pmatrix} \qquad (12)$$

or

$$\begin{pmatrix} a \\ b \\ c \end{pmatrix} \sqrt{n} = A \begin{pmatrix} a \\ b \\ c \end{pmatrix}, \qquad (13)$$

where $A = \begin{pmatrix} a_{11} & a_{12} & a_{13} \\ a_{21} & a_{22} & a_{23} \\ a_{31} & a_{32} & a_{33} \end{pmatrix}.$

---

[1] I would like to ask my readers unfamiliar with matrix notation and operations, determinants of matrices and systems of linear equations to read about them in any good secondary Algebra II text such as [D1] or in any introductory college text on linear algebra.

This situation, when the result $A \begin{pmatrix} a \\ b \\ c \end{pmatrix}$ of multiplication of the

matrix A by a vector $\begin{pmatrix} a \\ b \\ c \end{pmatrix}$ is collinear to $\begin{pmatrix} a \\ b \\ c \end{pmatrix}$, is very special. It is
studied in the area of mathematics called linear algebra. We need
some tools from linear algebra in order to study the equality (13)
and succeed with Grand Problem I.

3.2 Excursion in Linear Algebra

I will try to present here both an algebraic and a geometric view
of three-dimensional linear algebra. I will be very brief; I can't wait
to get back to geometry in general, and to our Grand Problem I in
particular!

Geometrically speaking, a *vector* $\vec{v}$ is an arrow from the origin.

By taking the Cartesian coordinates $\begin{pmatrix} a \\ b \\ c \end{pmatrix}$ of the endpoint of the ar-

row, we get an algebraic definition of a vector.

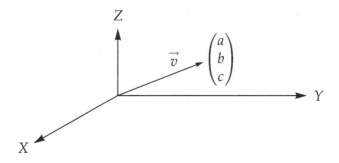

The set of all such vectors with the operations of addition and
multiplication by scalars is called the *vector space* $\mathbf{R}^3$. We would say
that a set of vectors $\vec{v_1}, \vec{v_2}, \vec{v_3}$ is *linearly independent* if the equation

$$\vec{v_1}x + \vec{v_2}y + \vec{v_3}z = \vec{0} \tag{14}$$

has a unique solution in real numbers:

$$x = y = z = 0.$$

Please note that for any set of vectors $\vec{v_1}, \vec{v_2}, \vec{v_3}, x = y = z = 0$ satisfies the equation (14). Linearly independent vectors $\vec{v_1}, \vec{v_2}, \vec{v_3}$ are the ones that allow *only* the trivial solution.

Let $\vec{v_1} = \begin{pmatrix} a_1 \\ b_1 \\ c_1 \end{pmatrix}, \vec{v_2} = \begin{pmatrix} a_2 \\ b_2 \\ c_2 \end{pmatrix}, \vec{v_3} = \begin{pmatrix} a_3 \\ b_3 \\ c_3 \end{pmatrix}$. Then the equation (14) would look like

$$\begin{pmatrix} a_1 \\ b_1 \\ c_1 \end{pmatrix}x + \begin{pmatrix} a_2 \\ b_2 \\ c_2 \end{pmatrix}y + \begin{pmatrix} a_3 \\ b_3 \\ c_3 \end{pmatrix}z = \begin{pmatrix} 0 \\ 0 \\ 0 \end{pmatrix},$$

or in coordinate form,

$$\begin{aligned} a_1x + a_2y + a_3z &= 0 \\ b_1x + b_2y + b_3z &= 0 \\ c_1x + c_2y + c_3z &= 0. \end{aligned} \tag{15}$$

Well, this is something familiar: a system of linear equations. We know that the system (15) has a unique solution if and only if the determinant of the system is not equal to zero:

$$\begin{vmatrix} a_1 & a_2 & a_3 \\ b_1 & b_2 & b_3 \\ c_1 & c_2 & c_3 \end{vmatrix} \neq 0.$$

Geometrically, three vectors $\vec{v_1}, \vec{v_2}, \vec{v_3}$ are linearly independent if and only if they do not lie in the same plane. And, of course, a set of vectors that is not linearly independent is called *linearly dependent*.

Please note that linear independence of vectors is defined here by the same equation as the integral independence of numbers in

Chapter 1. The only difference is in the set from which we take the values for our variables x, y, z. In Chapter 1, it was the set of integers. Here the set of scalars is the set of real numbers.

Problem 3.2.1. Give an example of a linearly independent set of three vectors and an example of a linearly dependent set.

I need to introduce you to one more fundamental notion of linear algebra. We say that the vectors $\vec{v_1}$, $\vec{v_2}$ and $\vec{v_3}$ *span* the entire vector space $\mathbf{R}^3$ if every vector $\vec{v}$ from $\mathbf{R}^3$ can be represented as a *linear combination* of the vectors $\vec{v_1}$, $\vec{v_2}$ and $\vec{v_3}$, i.e.,

$$\vec{v} = r_1\vec{v_1} + r_2\vec{v_2} + r_3\vec{v_3}$$

for some real numbers r_1, r_2, and r_3. For example:

Problem 3.2.2. Prove that the vectors $\begin{pmatrix} 1 \\ 0 \\ 0 \end{pmatrix}$, $\begin{pmatrix} 0 \\ 1 \\ 0 \end{pmatrix}$, $\begin{pmatrix} 0 \\ 0 \\ 1 \end{pmatrix}$ span the

vector space $\mathbf{R}^3$, i.e., every vector $\begin{pmatrix} a \\ b \\ c \end{pmatrix}$ of $\mathbf{R}^3$ can be represented

as

$$\begin{pmatrix} a \\ b \\ c \end{pmatrix} = r_1 \begin{pmatrix} 1 \\ 0 \\ 0 \end{pmatrix} + r_2 \begin{pmatrix} 0 \\ 1 \\ 0 \end{pmatrix} + r_3 \begin{pmatrix} 0 \\ 0 \\ 1 \end{pmatrix}$$

for some real numbers r_1, r_2, and r_3.

Solution. The required equality is satisfied by setting $r_1 = a$; $r_2 = b$; and $r_3 = c$. $\qquad\square$

A set of vectors $\vec{v_1}$, $\vec{v_2}$, $\vec{v_3}$ that is both linearly independent and spans the space $\mathbf{R}^3$ is called a *basis* of the space $\mathbf{R}^3$.

Proving the following important statement may make for an enjoyable pastime for my readers.

Problem 3.2.3. Prove that the following conditions on a set of *three* vectors $\vec{v_1}$, $\vec{v_2}$, $\vec{v_3}$ of $\mathbf{R}^3$ are equivalent (i.e., every condition implies every other):

1) vectors $\vec{v_1}, \vec{v_2}, \vec{v_3}$ are linearly independent;
2) vectors $\vec{v_1}, \vec{v_2}, \vec{v_3}$ span the space $\mathbf{R}^3$;
3) vectors $\vec{v_1}, \vec{v_2}, \vec{v_3}$ form a basis of $\mathbf{R}^3$.

One way to look at a 3×3 matrix A is as a source of a map, or linear transformation, of $\mathbf{R}^3$ into itself. Indeed, we can map every vector $\vec{v}$ into $A \cdot \vec{v}$:

$$\vec{v} \longmapsto A \cdot \vec{v}.$$

(A dot here denotes the matrix multiplication of a 3×3 matrix A by a 3×1 vector $\vec{v}$ with, of course, the 3×1 vector $A \cdot \vec{v}$ as the result.)

The image $A \cdot \vec{v}$ of a vector $\vec{v}$ can often be any vector at all. In fact, I'd like you to prove the following statement:

Problem 3.2.4. Let A be a 3×3 matrix with a non-zero determinant, and let $\vec{w}$ be an arbitrary vector. Then there is a vector $\vec{v}$ such that

$$A\vec{v} = \vec{w}.$$

As you can see, for matrices with a non-zero determinant, the image $A\vec{v}$ of the vector $\vec{v}$ can be any vector at all.

Yet matrices have their favorites: non-zero vectors $\vec{v}$ that do not change direction, but get stretched (i.e., get multiplied by a scalar when mapped by A):

$$\vec{v} \longmapsto A\vec{v} = \lambda\vec{v}.$$

These non-zero vectors, satisfying the equality

$$A\vec{v} = \lambda\vec{v} \tag{16}$$

for some real number λ, are called *characteristic vectors* of the matrix A corresponding to the *characteristic value* (i.e., coefficient of stretching) λ of A.

How do we find characteristic values λ and the corresponding characteristic vectors $\vec{v} = \begin{pmatrix} x \\ y \\ z \end{pmatrix}$ of a matrix $A = \begin{pmatrix} a_{11} & a_{12} & a_{13} \\ a_{21} & a_{22} & a_{23} \\ a_{31} & a_{32} & a_{33} \end{pmatrix}$?

No problem: according to (16),

$$\begin{pmatrix} a_{11} & a_{12} & a_{13} \\ a_{21} & a_{22} & a_{23} \\ a_{31} & a_{32} & a_{33} \end{pmatrix} \cdot \begin{pmatrix} x \\ y \\ z \end{pmatrix} = \lambda \begin{pmatrix} x \\ y \\ z \end{pmatrix}.$$

By performing the indicated multiplications in the matrix equation above and equating the corresponding components of the left and right sides, we get the following system of equations:

$$a_{11}x + a_{12}y + a_{13}z = \lambda x$$
$$a_{21}x + a_{22}y + a_{23}z = \lambda y$$
$$a_{31}x + a_{32}y + a_{33}z = \lambda z$$

or, equivalently,

$$\begin{array}{llll}
(a_{11} - \lambda)x & + \quad a_{12}y & + \quad a_{13}z & = \quad 0 \\
a_{21}x & + \quad (a_{22} - \lambda)y & + \quad a_{23}z & = \quad 0 \qquad (17) \\
a_{31}x & + \quad a_{32}y & + \quad (a_{33} - \lambda)z & = \quad 0.
\end{array}$$

Since we are looking for non-zero vectors $\begin{pmatrix} x \\ y \\ z \end{pmatrix}$, we need the system (17) to have at least one non-zero solution. (17) has a non-zero solution if and only if its determinant is equal to zero:

$$\begin{vmatrix} a_{11} - \lambda & a_{12} & a_{13} \\ a_{21} & a_{22} - \lambda & a_{23} \\ a_{31} & a_{32} & a_{33} - \lambda \end{vmatrix} = 0. \qquad (18)$$

If you compute the determinant in (18) (please do!), you get a cubic equation with respect to λ known as the *characteristic equation* of the matrix A. (We will have to talk a little more about cubic equations in Section 3.3 since they have surfaced here.)

Solutions of the characteristic equation are the characteristic values of A, and of course, by plugging a solution λ_0 of the characteristic equation in the system (17), we get the characteristic vectors of the matrix A corresponding to the characteristic value λ_0.

Please note (and prove!) the following statements.

Problem 3.2.5. For every characteristic value λ_0, we get at least one characteristic vector $\vec{v_0} = \begin{pmatrix} x \\ y \\ z \end{pmatrix}$ from (17).

In fact, the following important result has become a necessary part of every text on linear algebra.

Theorem 3.2.6. If the characteristic values λ_1, λ_2, and λ_3 of a 3×3 matrix A are distinct, then the corresponding characteristic vectors $\vec{v_1}$, $\vec{v_2}$, and $\vec{v_3}$ are linearly independent, and therefore form a basis of $\mathbf{R}^3$.

Proof. We are given precisely the following three equalities with non-zero vectors $\vec{v_1}$, $\vec{v_2}$, and $\vec{v_3}$:

$$\begin{aligned} A\vec{v_1} &= \lambda_1 \vec{v_1} \\ A\vec{v_2} &= \lambda_2 \vec{v_2} \\ A\vec{v_3} &= \lambda_3 \vec{v_3}. \end{aligned} \tag{19}$$

We need to prove that if

$$C_1 \vec{v_1} + C_2 \vec{v_2} + C_3 \vec{v_3} = \vec{0}, \tag{20}$$

where C_1, C_2, C_3 are real numbers, then necessarily $C_1 = C_2 = C_3 = 0$.

Indeed, assuming the equality (20) and multiplying it by A, we get:

$$A C_1 \vec{v_1} + A C_2 \vec{v_2} + A C_3 \vec{v_3} = \vec{0},$$

or

$$C_1 A\vec{v_1} + C_2 A\vec{v_2} + C_3 A\vec{v_3} = \vec{0}.$$

Now, using the equalities in (19), we come to

$$C_1\lambda_1\vec{v_1} + C_2\lambda_2\vec{v_2} + C_3\lambda_3\vec{v_3} = \vec{0}. \tag{21}$$

Multiplying (20) by λ_1 produces

$$C_1\lambda_1\vec{v_1} + C_2\lambda_1\vec{v_2} + C_3\lambda_1\vec{v_3} = \vec{0}. \tag{22}$$

By subtracting (22) from (21), we get rid of C_1:

$$C_2(\lambda_2 - \lambda_1)\vec{v_2} + C_3(\lambda_3 - \lambda_1)\vec{v_3} = \vec{0}. \tag{23}$$

Now we can repeat for the equality (23) the same procedure we applied to the equality (20), i.e., multiply (23) consecutively by A and λ_2 and subtract the second result from the first one:

$$C_2(\lambda_2 - \lambda_1)\lambda_2\vec{v_2} + C_3(\lambda_3 - \lambda_1)\lambda_3\vec{v_3} = \vec{0}$$
$$- \quad C_2(\lambda_2 - \lambda_1)\lambda_2\vec{v_2} + C_3(\lambda_3 - \lambda_1)\lambda_2\vec{v_3} = \vec{0}$$

$$C_3(\lambda_3 - \lambda_1)(\lambda_3 - \lambda_2)\vec{v_3} = \vec{0} \tag{24}$$

Since $(\lambda_3 - \lambda_1) \neq 0$, $(\lambda_3 - \lambda_2) \neq 0$, and $\vec{v_3} \neq 0$, the equality (24) implies $C_3 = 0$.

The coefficients C_1, C_2, and C_3 of equality (20) behave symmetrically. Therefore, by permuting them we can prove that $C_2 = 0$ and $C_1 = 0$ as well.

Thus, the vectors $\vec{v_1}$, $\vec{v_2}$, $\vec{v_3}$ are linearly independent, and due to the statement of Problem 3.2.3, they form a basis of $\mathbf{R}^3$.   □

You can easily prove (do!) the following statement:

Problem 3.2.7. Every multiple $\vec{v} = r\vec{v_1}$ of a characteristic vector $\vec{v_1}$ with non-zero real r is a characteristic vector that corresponds to the same characteristic value λ_1 as $\vec{v_1}$.

The converse of statement 3.2.7 is true only sometimes, but those situations are of prime interest to us.

Tool 3.2.8. Let the characteristic values λ_1, λ_2, and λ_3 of a 3×3 matrix A be distinct. If $\vec{v}$ and $\vec{v_1}$ are two characteristic vectors both corresponding to the same characteristic value λ_1, then

$$\vec{v} = r\vec{v_1}$$

for some real number r.

Proof of Tool 3.2.8 would serve you, my reader, as a final exam on this section of the book. It uses most of the ideas we discussed here and is similar to Theorem 3.2.6.

3.3 Excursion in Algebraic Equations

We will briefly (because Grand Problem I is still waiting for her solution!) discuss here cubic polynomials, since they surfaced in Section 3.2 as the characteristic polynomials of matrices. Most of these results are easily generalizable to algebraic equations of arbitrary degree. In fact, I will leave those generalizations for the "home entertainment" of my reader.

Tool 3.3.1. If $\{x_1, x_2, x_3\}$ is the set of roots (some roots may be equal in value) of a cubic equation

$$x^3 + a_1 x^2 + a_2 x + a_3 = 0, \tag{25}$$

then

$$x_1 + x_2 + x_3 = -a_1 \tag{26}$$

and

$$x_1 \cdot x_2 \cdot x_3 = -a_3. \tag{27}$$

Proof. We can decompose the given polynomial into factors as follows:

$$x^3 + a_1 x^2 + a_2 x + a_3 = (x - x_1)(x - x_2)(x - x_3).$$

By distributing on the right side, we get

$$x^3 + a_1 x^2 + a_2 x + a_3 = x^3 + (-x_1 - x_2 - x_3)x^2$$
$$+ (x_1 x_2 + x_1 x_3 + x_2 x_3)x + (-x_1 x_2 x_3). \quad (28)$$

Since (28) is true for every value of x, the corresponding coefficients on the left and right sides must be equal, and we obtain the required equalities (26) and (27). □

Tool 3.3.2. Let n, a_1, a_2, and a_3 be integers, and $n > 0$. If $\sqrt{n}$ is an irrational number and a root of the equation

$$a_0 x^3 + a_1 x^2 + a_2 x + a_3 = 0, \quad (29)$$

then $-\sqrt{n}$ is a root of the same equation.

Proof. Since $\sqrt{n}$ is a root of the equation (29), we have:

$$a_0 (\sqrt{n})^3 + a_1 (\sqrt{n})^2 + a_2 \sqrt{n} + a_3 = 0.$$

By distributing and combining terms, we get

$$(a_0 n + a_2)\sqrt{n} + (a_1 n + a_3) = 0, \quad (30)$$

which, due to Problem 1.3.10, implies (try to see how!) that

$$a_0 n + a_2 = 0 \quad \text{and} \quad a_1 n + a_3 = 0. \quad (31)$$

Now you can see that $-\sqrt{n}$ is a root of the equation (29) as well. Indeed, if we plug $-\sqrt{n}$ in for x in (29) and combine terms (please do), we get an expression differing from (30) by just one sign:

$$-(a_0 n + a_2)\sqrt{n} + (a_1 n + a_3) = 0,$$

which is an equality due to (31). □

When you read the statement of Tool 3.3.2 you may have wondered why I specified "an irrational" $\sqrt{n}$ for an integer n. What else can it be? Can it be a rational number and not an integer? No: in fact, *if n is an integer, $\sqrt{n}$ can only be an integer or else it is irrational!* It is a corollary (let us call it **Corollary 3.3.4**) of the following tool. I am leaving both proofs for you, my reader, to discover on your own.

Tool 3.3.3. Let $\dfrac{p}{q}$ be an irreducible fraction (p, q are integers, $q \neq 0$).

If $\dfrac{p}{q}$ is a root of the equation

$$a_0 x^n + a_1 x^{n-1} + \cdots + a_n = 0$$

with integer coefficients $a_0, a_1, \ldots, a_n$, then p is a divisor of a_n and q is a divisor of a_0.

Once you prove Tool 3.3.3 and its corollary 3.3.4, you can enjoy generalizing Tools 3.3.1 and 3.3.2.

Problem 3.3.5. Generalize Tool 3.3.1 to algebraic equations of degree n.

Problem 3.3.6. Generalize Tool 3.3.2 to algebraic equations of degree n.

4

How Does One Cut a Triangle? II

In Chapter 3, we compiled an impressive tool box. Will it enable us to complete the solution of Grand Problem I? There is one way to find out—try and see.

We know that every triangle can be cut into $1, 4, 9, 25, \ldots$ triangles congruent to each other (Chapter 2).

Assume that n is not a perfect square (i.e., $n \neq k^2$ for any integer k) and a triangle T is cut into n triangles congruent to each other. Repeating our unsuccessful attempt of Section 3.1, we assume that the angles of T are integrally independent. Then, by Tool 2.1, the congruent triangles of the partition are all similar to the triangle T, and the ratio of the corresponding sides is $1 : \sqrt{n}$. If the sides of n congruent triangles of the partition are a, b, and c, then, as in Section 3.1 (see Figure 3.1), we get the following system of equalities:

$$a\sqrt{n} = a_{11}a + a_{12}b + a_{13}c$$
$$b\sqrt{n} = a_{21}a + a_{22}b + a_{23}c$$
$$c\sqrt{n} = a_{31}a + a_{32}b + a_{33}c,$$

or in a matrix form

$$\vec{v_1}\sqrt{n} = A\vec{v_1},$$

A. Soifer, *How Does One Cut a Triangle?*, Second Edition, DOI 10.1007/978-0-387-74652-4_4,
© Alexander Soifer 2009

$$\text{where } \vec{v_1} = \begin{pmatrix} a \\ b \\ c \end{pmatrix} \text{ and } A = \begin{pmatrix} a_{11} \ a_{12} \ a_{13} \\ a_{21} \ a_{22} \ a_{23} \\ a_{31} \ a_{32} \ a_{33} \end{pmatrix}.$$

This means precisely that $\sqrt{n}$ is a characteristic value of the matrix A!

Since the entries a_{ij} of A are integers, all the coefficients of the characteristic equation E of the matrix A are integers. We know one root of E: $\sqrt{n}$. By Tool 3.3.2, the other root of E is $-\sqrt{n}$. The third root of E, thanks to Tool 3.3.1, must be an integer m.

We discovered that the matrix A has three distinct characteristic values: $\sqrt{n}$, $-\sqrt{n}$, and m. By Tool 3.2.8, every characteristic vector $\vec{v}$ corresponding to the characteristic value $\sqrt{n}$ is a scalar multiple of the characteristic vector $\vec{v_1}$:

$$\vec{v} = t\vec{v_1}. \tag{32}$$

Now let us consider the set $\mathbf{Q}[\sqrt{n}]$ of all numbers of the form $r_1 + r_2\sqrt{n}$, where r_1 and r_2 are rational numbers. This set has very impressive properties: if you add, subtract, multiply, or divide any two elements from $\mathbf{Q}[\sqrt{n}]$ (except division by 0, of course), you get an element from $\mathbf{Q}[\sqrt{n}]$ again.

In order to find a characteristic vector of the matrix A corresponding to the characteristic value $\sqrt{n}$, we have to solve the system (17). We do this by performing a number of additions, subtractions, multiplications, and divisions starting with the numbers a_{ij} and $\sqrt{n}$, i.e., the elements from $\mathbf{Q}[\sqrt{n}]$. In view of the properties of this set we observed in the previous paragraph, we can find a solution $\begin{pmatrix} x \\ y \\ z \end{pmatrix}$ of the system (17) such that all three numbers x, y, z are elements of the set $\mathbf{Q}[\sqrt{n}]$:

$$x = r_{11} + r_{12}\sqrt{n}$$
$$y = r_{21} + r_{22}\sqrt{n}$$
$$z = r_{31} + r_{32}\sqrt{n}.$$

This is the time to remember our "innocent" Problem 1.3.9: it shows that the numbers x, y, z are integrally dependent.

Now recall the equality (32): the characteristic vector $\vec{v} = \begin{pmatrix} x \\ y \\ z \end{pmatrix}$ is a scalar multiple of the characteristic vector $\vec{v_1} = \begin{pmatrix} a \\ b \\ c \end{pmatrix}$; i.e.,

$$\begin{pmatrix} x \\ y \\ z \end{pmatrix} = t \begin{pmatrix} a \\ b \\ c \end{pmatrix},$$

where t is a non-zero real number. The conclusion? *The components a, b, c of $\vec{v_1}$ are integrally dependent* because the components x, y, z of $\vec{v}$ are integrally dependent (show why this implication is true).

This may settle Grand Problem I. Indeed, let us take a triangle T satisfying the following two conditions:

(∗) The angles of T are integrally independent;
(∗∗) The sides of T are integrally independent.

Our reasoning from this chapter proves that such a triangle T cannot be cut into n triangles congruent to each other if n is not a perfect square.

We are done!

Are we?

Well, yes, if at least one triangle T satisfying the conditions (∗) and (∗∗) exists!

Can you think of one?

5

Excursion in Trigonometry

When I asked myself the last question of Chapter 4, I decided to try a few triangles and see whether they satisfied conditions (∗) and (∗∗). Would you believe it, the very first triangle I picked did the job! It was the triangle T with the sides $\sqrt{2}$, $\sqrt{3}$, $\sqrt{4}$ (see Figure 5.1).

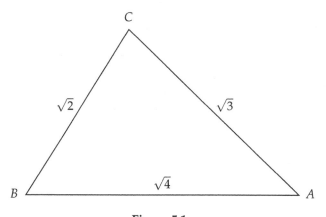

Figure 5.1

We already proved that the sides of T are integrally independent: that was exactly our Problem 1.3.1.

A. Soifer, *How Does One Cut a Triangle?*, Second Edition, DOI 10.1007/978-0-387-74652-4_5,   41
© Alexander Soifer 2009

We need to prove that the angles A, B, C of T are integrally independent as well, i.e., that the only integer solution of the Diophantine equation

$$Ax + By + Cz = 0 \tag{33}$$

is $x = y = z = 0$.

Let r denote a rational number and n a positive integer in the following table, which we will call Tool 5.1.

Tool 5.1.

ϕ	A	B
$\cos \phi$	$\dfrac{5}{4\sqrt{3}}$	$\dfrac{3}{4\sqrt{2}}$
$\sin \phi$	$\dfrac{\sqrt{23}}{4\sqrt{3}}$	$\dfrac{\sqrt{23}}{4\sqrt{2}}$
$\cos \phi \sin \phi$	$\dfrac{5\sqrt{23}}{48}$	$\dfrac{3\sqrt{23}}{32}$
$\cos^2 \phi$	$\dfrac{25}{48}$	$\dfrac{9}{32}$
$\cos 2n\phi$	r	r
$\sin 2n\phi$	$r\sqrt{23}$	$r\sqrt{23}$
$\cos(2n+1)\phi$	$r\sqrt{3}$	$r\sqrt{2}$
$\sin(2n+1)\phi$	$r\dfrac{\sqrt{23}}{\sqrt{3}}$	$r\dfrac{\sqrt{23}}{\sqrt{2}}$

Do you know how to use Tool 5.1? For example, if we need to find $\cos(2n+1)B$, we look at row $\cos(2n+1)\phi$ and column B. Their intersection shows us that $\cos(2n+1)B = r\sqrt{2}$ (where, as per our agreement, r stands for a rational number).

Tool 5.1 contains $8 \times 2 = 16$ entries, and therefore is a compact form for 16 statements. I would like you, my reader, to prove all these statements:

Problem 5.2. Prove the statements of the first row of Tool 5.1, i.e., $\cos A = \dfrac{5}{4\sqrt{3}}$ and $\cos B = \dfrac{3}{4\sqrt{2}}$. *Hint:* Use the Theorem of Cosines.

Problem 5.3. Prove the six statements of the second, third, and fourth rows of Tool 5.1.

Problem 5.4. Prove the eight statements of the last four rows of Tool 5.1. *Hint:* These statements can be proven all together by mathematical induction.

We need to fine-tune our Tool 5.1:

Tool 5.1′. In Tool 5.1, all four entries in rows $\cos 2n\phi$ and $\cos(2n+1)\phi$ are non-zero.

Proof. Let us prove, for example, that

$$\cos(2n+1)A \neq 0.$$

Indeed, if $\cos(2n+1)A = 0$, then $|\sin(2n+1)A| = 1$, but from Tool 5.1,

$$|\sin(2n+1)A| = r\frac{\sqrt{23}}{\sqrt{3}}.$$

We arrive at a contradiction because $1 \neq r\dfrac{\sqrt{23}}{\sqrt{3}}$. You can prove the three remaining statements similarly. ☐

Now with our custom-made Tools 5.1 and 5.1′, we are ready to look at the Diophantine equation (33). Assume that integers x, y, z, not all zero, form a solution of the equation (33), i.e.,

$$Ax + By + Cz = 0. \tag{33}$$

Since $C = \pi - A - B$, we get from (33) that

$$Ax_1 = -\pi z - By_1, \tag{34}$$

where $x_1 = x - z$ and $y_1 = y - z$.

Taking the cosine of both sides of equality (34), we obtain

$$|\cos x_1 A| = |\cos y_1 B|. \tag{35}$$

Please observe that x_1 and y_1 can not be both equal to zero, because then $\begin{cases} x - z = 0 \\ y - z = 0 \end{cases}$; therefore, $x = y = z$. From (33), $x = y = z = 0$ in contradiction with our assumption.

Assume that $x_1 \neq 0$ (the case of $y_1 \neq 0$ is similar). We can further assume that $x_1 > 0$ because replacing x_1 by $-x_1$ in (35) would not affect the equality. Now let x_2 be the *smallest* positive integer such that

$$|\cos x_2 A| = |\cos y_2 B| \tag{36}$$

for some integer y_2.

Our goal is, of course, to find a positive integer x_3 smaller than x_2 for which the equality (36) would work and thus reach a contradiction.

This is where we need (for the first time) our Tools 5.1 and 5.1'. They show that the equality (36) can take place *only* if both integers x_2 and y_2 are even! Let $x_2 = 2x_3$ and $y_2 = 2y_3$. By substituting them into (36), we get

$$|\cos 2x_3 A| = |\cos 2y_3 B|,$$

i.e.,

$$|2\cos^2 x_3 A - 1| = |2\cos^2 y_3 B - 1|. \tag{37}$$

Removing the absolute value signs in (37) requires us to consider two cases.

Case 1. Suppose $2\cos^2 x_3 A - 1 = 1 - 2\cos^2 y_3 B$.

Then $\cos^2 x_3 A = \sin^2 y_3 B$, i.e.,

$$|\cos x_3 A| = |\sin y_3 B| .$$

The tools 5.1 and 5.1' show that the last equality cannot take place!

Case 2. Suppose $2\cos^2 x_3 A - 1 = 2\cos^2 y_3 B - 1$.
  Then

$$|\cos x_3 A| = |\cos y_3 B| ,$$

and we indeed found an integer x_3 such that $0 < x_3 < x_2$ satisfying (36), which contradicts the minimality of x_2.

The angles A, B, and C of the triangle T are integrally independent. We have reached the summit of Grand Problem I!   □

Why did it happen that the first triangle I picked satisfied both conditions of integral independence of sides and integral independence of angles? In part, it is, of course, my experience. More importantly, however, is the fact that perhaps "most" triangles satisfy these two conditions (∗) and (∗∗).
  I would like to pose the following question for a prepared reader:

Problem 5.5. How large is the set of all non-similar triangles satisfying the conditions (∗) and (∗∗)? How large is the set of all non-similar triangles which do not satisfy at least one of the conditions (∗) or (∗∗)?

I created (and solved) Grand Problem I in 1970. During the two decades that followed, I offered it to many young and not-so-young mathematicians. The only solution came in 1989 from the brilliant problem solver and my friend, Semion Slobodnik (see more about him in Section 8.7).
  The early stages of his proof coincide with my own: he proves Tool 2.1. Then our proofs part ways. I, an algebraist, finish up with an algebraic Chapter 4. He, an analyst by training, beautifully uses mathematical analysis to prove that if every triangle can be cut into k congruent triangles, then $\sqrt{k}$ is an integer. Please, do read it in Appendix A at the end of the book.

6

Is There Anything Beyond the Solution?

When I finished writing down the solution of Grand Problem I, I did not want to think about anything else. It was time to celebrate. I took a nice walk around Quail Lake.

When I came back, I knew what was beyond the successful solution to a problem: new problems that stem from the solved one! I felt as if I were at a fork. My road was splitting. Here are your options, the roads to walk and enjoy (many of them, but not all, I have walked myself):

Problem 6.1. For *each* triangle T, find all positive integers n such that the triangle T can be cut into n triangles *similar to* T.

Problem 6.2. For *each* triangle T, find all positive integers n such that the triangle T can be cut into n triangles *similar to each other*.

Problem 6.3. (I. M. Yaglom [Y1]) For each triangle T, find a partition of T into triangles T_i similar to T such that all triangles T_i of the partition are incongruent. Classify such partitions.

The following problem I solved using an enjoyable but elaborate geometric induction:

Problem 6.4. Prove that if the angles of a triangle T are integrally independent and the sides of T are integrally independent, then for

A. Soifer, *How Does One Cut a Triangle?*, Second Edition, DOI 10.1007/978-0-387-74652-4_6,   47
© Alexander Soifer 2009

any integer k, *the only way* to cut T into k^2 triangles congruent to each other is the way presented in Figure 2.2.

Problem 6.5. For *each* triangle T, find all positive integers n such that T can be cut into n triangles congruent to each other.

For each T and n find the number of distinct partitions of T into n congruent triangles.

The last problem is open and very difficult. Do write to me if you discover any solutions, even partial ones. In fact, *Paul Erdős offered a twenty-five-dollar prize for the first solution to the following two related problems.*

Problem 6.6. (P. Erdős, $25 problem). Find (and classify) *all* triangles that can *only* be cut into n^2 congruent triangles for any integer n.

Problem 6.7. (P. Erdős, $25 problem). Find all positive integers n such that *at least one* triangle can be cut into n triangles congruent to each other.

Problem 6.7 is open, but I would like to share with you a couple of relevant results.

Problem 6.8. Prove that for any positive integer n there is a triangle that can be cut into $2n^2$, $3n^2$, and $6n^2$ triangles congruent to each other.

Problem 6.9. (XV Russian Mathematical Olympiad, 1989) Prove that for every n that is the sum of two perfect squares (of integers), there is a triangle that can be cut into n triangles congruent to each other.

Well, now let us "cool down." The following few problems are nice, short, and relatively easy. Work on them on your own.

Problem 6.10. Given an obtuse triangle T, find the minimal number of acute triangles T can be cut into.

Hint: Grand Problem II had a nice answer: all positive integers, except the first three primes 2, 3, 5. Interestingly, this problem's answer is the fourth prime: 7.

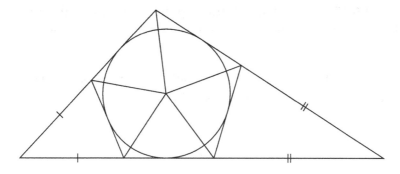

Figure 6.1

A way to cut an obtuse triangle T into seven acute triangles is demonstrated in Figure 6.1.

Prove that the method of Figure 6.1 always works. Then there is "only" left to prove that it is impossible to cut T into less than seven acute triangles.

Problem 6.11. Given an acute triangle T, find the minimal number of obtuse triangles T can be cut into.

Problem 6.12. Given an obtuse triangle T, what is the minimal number (greater than 1) of obtuse triangles T can be cut into?

Problem 6.13. Given an acute triangle T, what is the minimal number (greater than 1) of acute triangles T can be cut into?

Problem 6.14. Given a right triangle T, what is the minimal number of

a) acute triangles T can be cut into?
b) obtuse triangles T can be cut into?

The last problem serves as a nice illustration that a right triangle sometimes acts like an obtuse triangle, and sometimes like an acute one.

Finally, let me offer you a problem that I was lucky to create in 1966 when I was a high school senior.

Problem 6.15. Partition an arbitrary triangle by six straight cuts into such parts from which one can put together seven congruent triangles.

I included the solution of this problem in my book, *Mathematics as Problem Solving* ([S1], Problem 4.5.3).

7

Pursuit of the Best Result

7.1 Birth of a Problem

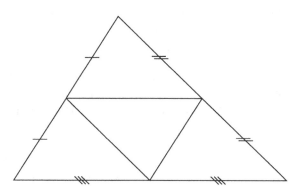

Figure 7.1

What do you think about when you look at Figure 7.1? Me, I am not particularly in love with this textbook illustration. And yet, it was the inspiration of Grand Problems I and II—and not only them!

In August 1987, I taught a problem solving course for gifted high school students from several countries. For their exam I wanted to create an easy but original Pigeonhole Principle problem of geomet-

A. Soifer, *How Does One Cut a Triangle?*, Second Edition, DOI 10.1007/978-0-387-74652-4_7,
© Alexander Soifer 2009

ric flavor. In case you are not familiar with the Pigeonhole Principle, here it is:

If $kn + 1$ pigeons (k, n are positive integers) sit in n pigeonholes, then at least one of the holes contains at least $k + 1$ pigeons.

For the proof of the principle as well as a number of exciting problems, solved and unsolved, I refer my readers to Sections 1.3 and 1.4 of my book, *Mathematics as Problem Solving* [S1].

Figure 2.1 gave me the initial problem. This is what I thought about when I looked at Figure 7.1.

Problem 7.1.1. Given nine points in a triangle (i.e., in the interior or on the boundary) of area 1, prove that three of them form a triangle of area not exceeding $1/4$.

Solution. Midlines partition the given triangle into four congruent triangles of area $1/4$ (Figure 7.1). These congruent triangles are our pigeonholes. The given points are our pigeons. Now nine pigeons are sitting in four pigeonholes. Since $9 = 2 \cdot 4 + 1$, there is at least one pigeonhole containing at least three pigeons.   □

If you, my reader, feel that nine points are excessive to guarantee the result in Problem 7.1.1, you are quite right. But in order to prove the following stronger statement, we need to allow the pigeonholes to be different in size and shape.

Problem 7.1.2. Given seven points in a triangle of area 1, prove that three of them form a triangle of area not exceeding $1/4$.

First Solution. Since $7 = 2 \cdot 3 + 1$, it would be nice to have three pigeonholes; then at least one of them would contain at least three pigeons, and we would be done!

Let us draw only two midlines in the given triangle (Figure 7.2). We get three pigeonholes. At least one of them contains at least three pigeons. If one of the triangles contains three given points, we are done.

If the parallelogram contains three given points, then all we have left to prove is a simple tool:

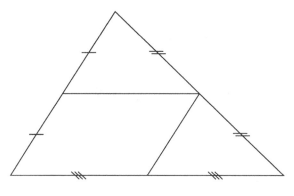

Figure 7.2

Tool 7.1.3. The maximum area of a triangle inscribed in a parallelogram of area $1/2$ is equal to $1/4$.

I leave the proof of this tool to my readers.

Second Solution. As we sensed in the beginning of the first solution, we ought to partition the given triangle ABC into three pigeonholes. Let us do this by drawing two lines parallel to the base AC of the triangle (Figure 7.3): a midline MN and a line PQ equidistant from AC and MN.

By the Pigeonhole Principle, at least three of the seven given points will be in one pigeonhole. One of the pigeonholes is the triangle MBN of area $1/4$. Two others can easily be embedded in parallelograms (see Figure 7.3) of area not exceeding $1/2$. Therefore, any triangle inscribed in them has area not exceeding $1/4$ (see Tool 7.1.3). □

It is time to invite you, my reader, to prove that in fact six points are enough:

Problem 7.1.4. Given six points in a triangle of area 1, prove that three of them form a triangle of area not exceeding $1/4$.

In our aspiration for the optimal result, we reduced the number of given points from nine to seven to six. How far can we go?

Well, not very far:

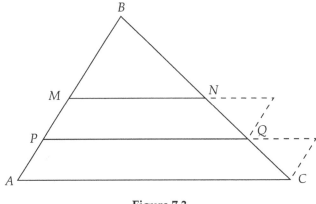

Figure 7.3

Problem 7.1.5. In any triangle of area 1, there are four points such that every three of them form a triangle of area greater than 1/4.

Solution. It suffices to note that the three vertices of the given triangle plus its center of mass (i.e., the point of intersection of medians) form the required four points (Figure 7.4).

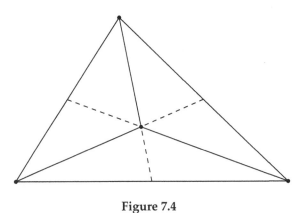

Figure 7.4

Indeed, the smallest area of a triangle formed by three of the selected four points is equal to 1/3.   □

Thus, with a probability of 50%, we can guess the minimal number of points that will guarantee the result: five or six! I hope by now you have proven (Problem 7.1.4) that six points are enough. In the next section we will look at the optimal result.

7.2 The Optimal Result

Problem 7.2.1. (The Five-Point Problem) Of any five points in a triangle of area 1, there are three points that form a triangle of area not exceeding $1/4$.

Proof. Assume that five points are given in a triangle of area 1 and that no three points form a triangle of area not exceeding $1/4$. We will try to see where these five points can be situated.

Once again, let us draw all three midlines (Figure 7.5 (a)), and each pair (Figures 7.5 (b), (c), (d)).

Let S_0 denote the midlines triangle in Figure 7.5 (a), including its boundary; and let S_1, S_2, and S_3 denote the parallelograms in Figures 7.5 (b), (c), and (d), including their boundaries. We will use the symbol $m(S_i)$ for the number of the given points contained in S_i. We know that for any figure S_i defined above, $m(S_i) \leq 5$ since there are exactly five points in the given triangle.

In fact,

$$m(S_1) \leq 2$$
$$m(S_2) \leq 2 \tag{38}$$
$$m(S_3) \leq 2,$$

because if $m(S_i) \geq 3$ for some $i = 1, 2, 3$, the parallelogram S_i contains at least three given points. Then, due to Tool 7.1.3, these three points form a triangle of area not exceeding $1/4$, in contradiction to our assumption.

Now observe that if we put Figures 7.5 (b), (c), and (d) on top of Figure 7.5 (a) so that the given triangles on all four figures coincide, the parallelograms S_1, S_2, and S_3 will completely cover the given triangle once, and the triangle S_0 will be covered two extra times.

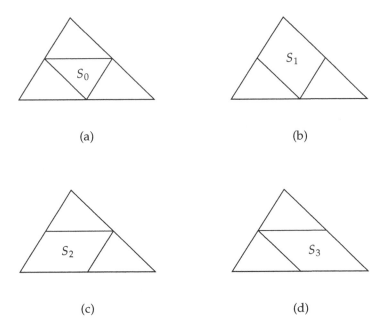

(a)   (b)

(c)   (d)

Figure 7.5

This observation has a neat consequence:

$$m(S_1) + m(S_2) + m(S_3) = 5 + 2m(S_0).$$

In other words, in view of (38),

$$2 + 2 + 2 \geq 5 + 2m(S_0)$$

and therefore

$$m(S_0) \leq \frac{1}{2}.$$

But $m(S_0)$ is a whole number, therefore,

$$m(S_0) = 0$$

and no given points are in the middle triangle of Figure 7.5 (a)!

Then where *are* they?

They are in the three corner triangles of Figure 7.5 (a) and distributed 1–2–2 (the original assumption prevents the corner triangles from containing three or more given points). Without loss of generality, let us assume that the top triangle of Figure 7.5 (a) contains exactly one given point.

Let us partition the given triangle into 16 little congruent triangles by drawing straight lines parallel to the sides at equal intervals (Figure 7.6), and mark in black the triangle S_0 (which we know contains no given points).

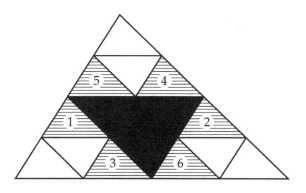

Figure 7.6

Let i_1 be the number of given points contained in the little triangle marked by 1 (see Figure 7.6). We define the numbers $i_2, i_3, \ldots, i_6$ similarly.

You may recall the second solution to Problem 7.1.2. Since, as we noticed above, the triangle MBN (see Figure 7.3) contains exactly one given point, the region $PMNQ$ must contain exactly 2 given points (can you figure out why?). Now look at Figure 7.6. All the given points in the region $PMNQ$ must lie in the triangles marked 1 and 2. Therefore,

$$i_1 + i_2 = 2. \tag{39}$$

Similarly, we can show (do!) that

$$i_3 + i_4 \geq 1$$

and   (40)

$$i_5 + i_6 \geq 1.$$

Now I would like to introduce the following sums:

$$\Sigma_1 = i_1 + i_5 + i_4; \quad \Sigma_3 = i_2 + i_4 + i_5; \quad \Sigma_5 = i_3 + i_6 + i_1$$
$$\Sigma_2 = i_1 + i_5 + i_3; \quad \Sigma_4 = i_2 + i_4 + i_6; \quad \Sigma_6 = i_3 + i_6 + i_2$$ (41)

Why did I define them? It is *my* solution and, therefore, I can do whatever I feel like in it! To be a bit more serious, give me a little time, and I will show you that it works.

For starters, we can prove the following inequality using (41), (39), and (40):

$$\Sigma_1 + \Sigma_2 + \Sigma_3 + \Sigma_4 + \Sigma_5 + \Sigma_6$$
$$= 3(i_1 + i_2 + i_3 + i_4 + i_5 + i_6) \geq 12.$$ (42)

On the other hand, we can prove that

$$\text{for every } j = 1, 2, \ldots, 6, \quad \Sigma_j \leq 2.$$ (43)

Indeed, assume the opposite: let, for example, $\Sigma_3 \geq 3$. Since $\Sigma_3 = i_2 + i_4 + i_5$, the triangles 2, 4, and 5 contain at least 3 given points, say v_1, v_2, v_3. But *we can surround all three triangles 2, 4, 5 by one parallelogram of area* $1/2$ (see Figure 7.7).

Due to Tool 7.1.3, the points v_1, v_2, v_3 form a triangle of area at most $1/4$ in contradiction to the original assumption.

Now you know the secret behind my sums Σ_j in (41): just as we showed for Σ_3, each Σ_j *counts the given points in the region embeddable in a parallelogram of area* $1/2$, *and, therefore, each* $\Sigma_j \leq 2$! The inequalities of (43) are proven.

The inequalities (42) and (43) barely coexist; they can be satisfied only by

$$\Sigma_1 = \Sigma_2 = \Sigma_3 = \Sigma_4 = \Sigma_5 = \Sigma_6 = 2.$$

But $\Sigma_1 = \Sigma_2$ means, by (41), that

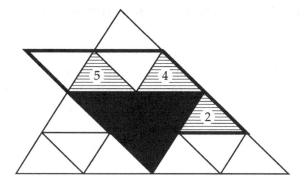

Figure 7.7

$$i_1 + i_5 + i_4 = i_1 + i_5 + i_3,$$

i.e.,

$$i_4 = i_3.$$

But by (40), $i_3 + i_4 \geq 1$. Therefore, since i_3 and i_4 are whole numbers, we get

$$i_3 + i_4 \geq 2. \tag{44}$$

Similarly, $\Sigma_3 = \Sigma_4$ implies $i_5 = i_6$, and therefore

$$i_5 + i_6 \geq 2. \tag{45}$$

Now, let us count the given points. We know that there are no given points in the triangle S_0 (Figure 7.5 (a)). Therefore, triangles 1, 2, 3, 4, 5, and 6 of Figure 7.6 have no given points in common. And yet, due to (39), (44), and (45),

$$i_1 + i_2 + i_3 + i_4 + i_5 + i_6 = 6,$$

i.e., in just triangles $1, 2, \ldots, 6$ *we have accounted for six given points, whereas we only had five points to begin with!*

This contradiction shows that our assumption that no three given points form a triangle of area $1/4$ or less is false. □

7.3 Second Solution to the Five-Point Problem

Once I created this problem, I kept it a secret until April 1988, when I presented it at the Fifth Annual Colorado Mathematical Olympiad. I also sent it to a number of young and not-so-young mathematicians, and finally offered it (with a prize for a solution!) to my audience at the Sixth International Congress on Mathematical Education in August 1988 in Budapest. I received solutions from four very interesting mathematicians. Only one solution resembled my solution in Section 7.2. Its author, Renny Thoms from Hermiston, Oregon, or as he put it, "from the middle of nowhere, Oregon," was my student in the summer of 1987, and is now a freshman at Stanford University.[1]

Three other solutions were based on similar ideas that were totally different from my solution. They came from my summer 1987 and 1988 student, Royce Peng,[2] now a senior at Rancho Palos Verdes High School in California and a candidate for the American team for the International Mathematical Olympiad; Dr. Janoŝ Pataki from Budapest, a coach of the Hungarian team for the International Mathematical Olympiad; and the famous Russian mathematician Vladimir Boltyanski.

The following solution is essentially due to Royce Peng—I just simplified and shortened it. But first, another tool:

Tool 7.3.1. The locus of all vertices B of triangles ABC with the given base $\overline{AC}$ and the given area S is a pair of straight lines L_1, L_2 parallel to $\overline{AC}$ lying on either side of $\overline{AC}$ at the distance $\dfrac{2S}{|AC|}$ (see Figure 7.8). Moreover, if the area of a triangle AXC is less than S, then the point X lies between the lines L_1 and L_2. If the area of a triangle AYC is greater than S, then the point Y lies outside of the strip bounded by the lines L_1 and L_2.

---

[1] Renny Thoms has since obtained a masters degree from Rutgers University. He is now a software developer.

[2] Royce Peng has since obtained degrees from the University of Southern California and Harvard.

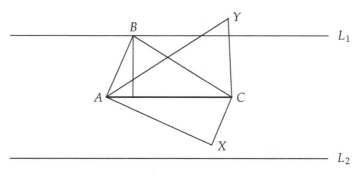

Figure 7.8

I will refer the readers who need assistance with proving Tool 7.3.1 or would like to have more problems on loci to my *Mathematics as Problem Solving* [S1].

Now we are ready for the solution. Midlines (again!) partition the given triangle into four pigeonholes (Figure 7.1). At least one of the pigeonholes must contain at least two of the five given points.

If the midlines triangle MNK contains two given points v_1 and v_2 (Figure 7.9), then we are done.

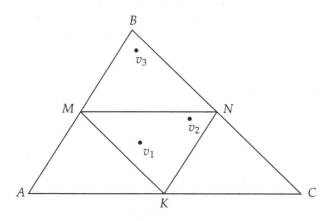

Figure 7.9

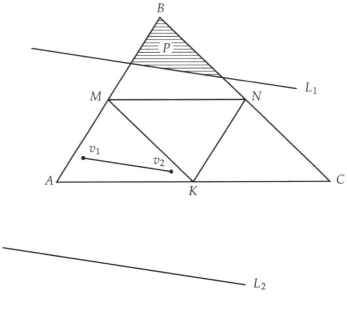

Figure 7.10

Indeed, one of the corner triangles, say MBN, must contain at least one of the three remaining given points v_3, and we can surround the three points v_1, v_2, and v_3 by a parallelogram $MBNK$ of area $1/2$. By Tool 7.1.3, this guarantees that the area of the triangle $v_1v_2v_3$ is at most $1/4$.

Assume now that one of the corner triangles, say AMK, contains at least two given points v_1 and v_2 (Figure 7.10).

Now we would like to draw the locus of all vertices K of triangles v_1Kv_2 of area $1/4$. Tool 7.3.1 tells us that this locus is a pair of lines L_1 and L_2 parallel to $\overline{v_1v_2}$. But where can L_1 and L_2 lie? Can they intersect MN, for example?

Since each of the triangles v_1Mv_2, v_1Nv_2, v_1Kv_2, and v_1Av_2 is contained in the parallelogram $AMNK$ of area $1/2$, by Tool 7.1.3, each of these triangles has area of at most $1/4$.

By the second sentence of Tool 7.3.1, we can conclude that the points A, M, N, and K all lie between or on the lines L_1 and L_2. What part of triangle ABC can possibly lie outside of the strip bounded by the lines L_1 and L_2? *Only* a piece P of the triangle MBN *or* the triangle NCK! Can you prove that pieces of *both* triangles MBN and NCK cannot be outside of the strip?

Now we are done:

Either one more of the given points v_3 lies between the lines L_1 and L_2 (or on L_1 or L_2), and the area of the triangle $v_1 v_2 v_3$ is $1/4$ or less (can you tell why?), *or else* all three remaining given points v_3, v_4, and v_5 lie in the piece P of the triangle MBN (or triangle NCK), and the area of triangle $v_3 v_4 v_5$ is surely less than or equal to the area of triangle MBN, which is exactly $1/4$!   □

I received an alternate version of my proof of the Five-Point Problem (Problem 7.2.1). It came from the wonderful problem solver, Professor Cecil Rousseau from Memphis State University, who for a number of years was the coach of the American team for the International Mathematical Olympiad. I hope you will enjoy this elegant version as much as I have. You can find it in Appendix B at the end of the book.

8

Convex Figures and the Function $S(F)$

*Science is always wrong. It never solves a problem
without creating ten more.*

—*George Bernard Shaw*

8.1 Creating a Function

A perfect result is achieved, a series of problems is solved. Now
what? Well, a good solution to a good problem generates many new
questions. That is why mathematics will always be alive!

We have found an interesting property of the triangle: in any set
of five points in a triangle of area S there are three points that form a
triangle of area at most $S/4$. Moreover, five is the minimal number
of points in a triangle that guarantees this result.

Why did we study a triangle alone? Oh yes, this book's title is
How Does One Cut a Triangle? —Not a good enough reason!

Notation. For a geometric figure F, $|F|$ will denote the *area* of F.
For example, $|ABC|$ denotes the area of the triangle ABC; $|ABCD|$
stands for the area of the quadrilateral $ABCD$.

Definition 8.1.1. Given a figure F, let $S(F)$ denote the minimal positive integer n such that among any n points located inside or on the
boundary of F there are always three points which form a triangle
of area at most $|F|/4$.

In this terminology, the result we obtained in Chapter 7 can be
stated as follows:

A. Soifer, *How Does One Cut a Triangle?*, Second Edition, DOI 10.1007/978-0-387-74652-4_8,
© Alexander Soifer 2009

$$\text{For any triangle } \Delta, \quad S(\Delta) = 5.$$

To begin with, what are the values of $S(F)$ for some familiar figures F?

Here is an easy problem for your entertainment:

Problem 8.1.2. Prove that for any parallelogram P,

$$S(P) = 5.$$

Now we have a new problem, therefore we are alive! And the problem is this: What are all possible values of our newly introduced function $S(F)$? Can the function $S(F)$ help us classify geometric figures?

We will only discuss bounded figures, i.e., the figures that can be contained by a square. Thus, from now on the word "figure" means "bounded figure."

Problem 8.1.3. $S(F)$ exists for every figure F.

Proof. Let F be a figure of area $|F|$. There is a square L completely containing F. We can divide L into slices of area $|F|/2$ or less (Figure 8.1).

If the number of slices is m, then any $2m + 1$ points of figure F contain three points that form a triangle of area at most $|F|/4$. Indeed, by the Pigeonhole Principle (the slices are our pigeonholes and the points are our pigeons), there is a slice which contains at least three points v_1, v_2, v_3. By Tool 7.1.3, these three points form a triangle of area at most half the area of a slice.

To complete the proof, we need to know what positive integers are. Do you know?

Yes, you are right: they are elements of the set

$$N = \{1, 2, \ldots, n, \ldots\}.$$

But at the end of the nineteenth century, mathematicians discovered that this "naive" definition proved to be not good enough.

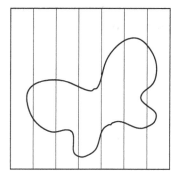

Figure 8.1

Many areas of mathematics were put on an axiomatic foundation. The Italian mathematician Giuseppe Peano (1858–1932) created a system of axioms for the set of positive integers that became universally accepted. For our purposes we need the *axiom of minimal element:*

Any non-empty subset N_1 of the set of positive integers N contains a minimal (smallest) element.

Let us apply this axiom to our situation. Denote by N_1 the subset of N consisting of positive integers n such that any set of n points of figure F contains three points v_1, v_2, v_3 that form a triangle of area $|F|/4$ or less. This set N_1 is not empty; as we proved above, it contains the number $2m + 1$. Therefore, by the axiom of minimal element, N_1 contains a smallest positive integer n_1.

By the definition of the function $\mathbf{S}(F)$,

$$\mathbf{S}(F) = n_1. \qquad \square$$

Problem 8.1.4. The function $\mathbf{S}(F)$ is unbounded. For any positive integer m, there is a figure F such that

$$\mathbf{S}(F) > m.$$

Proof Outline. There exists a large enough regular m-gon such that the area of a triangle with vertices at three consecutive vertices v_1, v_2, v_3 of the m-gon is greater than $1/4$ (Figure 8.2).

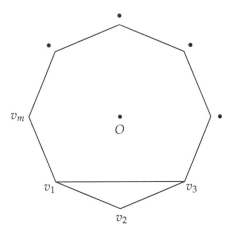

Figure 8.2

Now we can draw the required figure F as a snowflake with the center in O and m radii going from O to each vertex of the m-gon.

We draw radii thin enough so that the area of the figure F is equal to 1 (see Figure 8.3).

Now we pick a set of m points in figure F: the endpoints of the radii $v_1, v_2, \ldots, v_m$. The smallest area of a triangle with vertices in these m points is equal to the area of the triangle $v_1 v_2 v_3$ (Figure 8.2), which is greater than $1/4$; therefore, m points are not enough to guarantee that three of them form a triangle of area $1/4$ or less. I.e.,

$$S(F) > m.$$

It so nicely happened that the very next day after I finished the first rough draft of this book, I was already lecturing out of it. John Benson and his colleagues had invited me to share my problems with their talented young mathematicians at Evanston Township High School (north of Chicago) on December 1st and 2nd, 1989. It

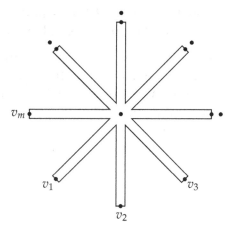

Figure 8.3

was a wonderful exchange of ideas. I taught and I learned. David Revelle, then a junior, created a tire to use in place of my snowflake (Figure 8.3′):

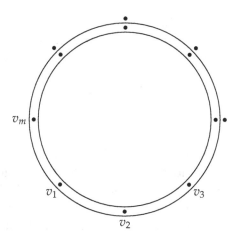

Figure 8.3′

It works: we can make a studded tire (studded by the points $v_1, v_2, \ldots, v_m$) thin enough that its area is equal to one.   □

The result we have just proved is interesting, but it really does not contribute to our knowledge of geometric figures: $S(F)$ can be any integer large enough. In our proof we used the construction of an essentially non-convex snowflake. It appears that we could not have done it with a convex figure F. Let us therefore study the function $S(F)$ on convex figures F. $S(F)$ should not be unbounded in this case, which might prove to be an important classification of convex figures F by the values $S(F)$.

8.2 Study of Convex Figures: The Upper Bound of $S(F)$

Without making our research any less attractive, we will limit our consideration to *convex figures*.

Definition 8.2.1. A geometric figure F is called *convex* if for every two points v_1 and v_2 of F, every point of the segment $\overline{v_1 v_2}$ is in F.

Figures 8.2 and 8.3 of the previous section give us examples of convex and non-convex figures, respectively. There is another way to describe convex figures.

Definition 8.2.1′. A plane geometric figure F is called convex if it can be obtained as the intersection of a finite or infinite set of half-planes.

A convex figure F can be obtained as the intersection of all half-planes that contain F and are bounded by the tangent lines of F if tangent lines exist at every boundary point of F (Figure 8.4). If the set of half-planes is finite, we certainly get a convex polygon F (Figure 8.5). Figure 8.6 gives you another example of a non-convex figure.

From now on the word "figure" means "bounded convex figure."

Before we get down to business, let me introduce to you here, without proofs, a few tools from mathematical analysis. Building

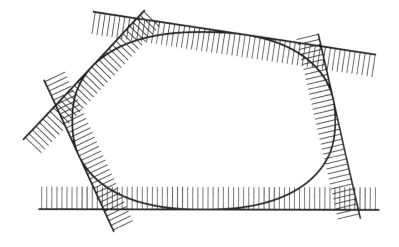

Figure 8.4

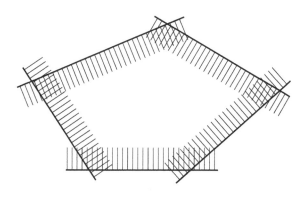

Figure 8.5

up (i.e., proving) this machinery here would take us too far from our research and for too long.

Tool 8.2.2. (Intermediate Value Theorem) Let $f(x)$ be a continuous function on a segment $[a, b]$. If c is any real number between $f(a)$ and $f(b)$, then there exists a point x_0 on the segment $[a, b]$ such that $f(x_0) = c$.

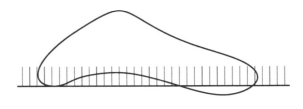

Figure 8.6

In other words, *between any two of its values a continuous function takes on all intermediate values.*

Tool 8.2.3. Given a geometric figure F, a horizontal axis, and a vertical line L (Figure 8.7). The area $f(x)$ of the part of figure F located to the left of L can be viewed as a function of the x-coordinate of the intersection of L with the axis.

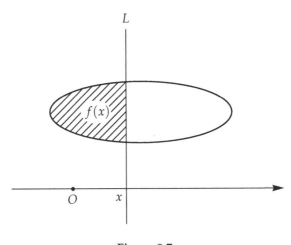

Figure 8.7

This function $y = f(x)$ is continuous.

Tool 8.2.4. Given a figure F, a horizontal axis with origin O, and a line L through O with unit vector $\vec{v}$ attached to L (Figure 8.8).

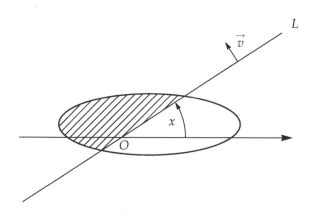

Figure 8.8

The area $f(x)$ of the part of figure F located in the half-plane that is bounded by L and contains $\vec{v}$ can be viewed as a function of the angle x between L and the axis. This function is continuous.

Tool 8.2.5. For any convex figure F there exists:

a) an inscribed triangle T of maximum area;
b) an inscribed parallelogram P of maximum area;
c) an inscribed ellipse E of maximum area.

Now it is time to work. Prove the following tool:

Tool 8.2.6. In the setting of Tool 8.2.5, the vertices of the optimal triangle T and parallelogram P lie on the boundary of the figure F.

I would really like to show you how Tools 8.2.2, 8.2.3, and 8.2.4 work together to produce the following three results, but I do not wish to deny you the pleasure of discovering it on your own!

Tool 8.2.7. For any geometric figure F and any point p there exists a line L through p that divides F into two pieces of equal area.

Tool 8.2.8. For any point p on the boundary of a convex figure F there exists a line L through p such that the entire figure F lies in one of the two half-planes determined by L (we include L in both half-planes).

We will call L a *supporting line* for F at p.

Please note that this notion of a supporting line, which is convenient to our purposes, is broader than the more familiar notion of a tangent line. It allows in some cases more than one and, therefore, an infinite number of supporting lines for F, all at the same point p. Can you think of an example?

Tool 8.2.9. For any geometric figure F and any line L, there are exactly two supporting lines for F parallel to L.

The next result is clearly inspired by Problem 7.1.1.

Problem 8.2.10. For any convex figure F,

$$S(F) \leq 9.$$

Solution. Of course, we cannot draw midlines as we did in the solution of problem 7.1.1 (to begin with, we don't know what "midline" would mean for an arbitrary figure F!). But we can divide F into four pieces of equal area.

Here is one of many ways to do it. We take a vertical line L to the left of figure F and start moving it to the right (Figure 8.9). The area S_L of the part of F lying to the left of L changes continuously from 0 to the entire area $|F|$ of F. Therefore, by Tools 8.2.3 and 8.2.2, at some point $S_{L_1} = \dfrac{1}{4}|F|$; at some other point $S_{L_2} = \dfrac{1}{2}|F|$; and at yet another point $S_{L_3} = \dfrac{3}{4}|F|$. We are done dividing F.

Now we can repeat the reasoning of the solution to Problem 7.1.1. Since nine given points (pigeons) are located in four pieces

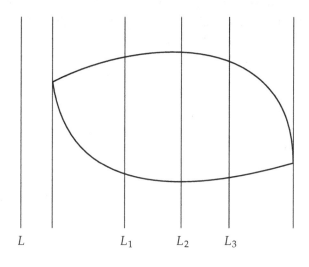

Figure 8.9

of figure F (pigeonholes), there is a piece containing at least three of the given points v_1, v_2, v_3. These points form a triangle of area at most $|F|/4$ because the entire triangle $v_1 v_2 v_3$ is contained in the piece of F of area $|F|/4$ (remember, F is convex and, therefore, each of our four pigeonholes is convex!).   □

We can do better than that. My proof of the following result is based on the idea that Gideon Yaffe, a senior from the Colorado Springs School, used in his solution to Problem 7.1.2 during the Fifth Annual Colorado Mathematical Olympiad in 1988. He won one of two first prizes in our Olympiad and is now a sophomore at Harvard majoring in mathematics and... drama![1]

Problem 8.2.11. For any convex figure F,

$$S(F) \leq 6.$$

---

[1] Gideon Yaffe is now a professor of philosophy and law at the University of Southern California.

Solution. Assume that points $v_1, v_2, \ldots, v_6$ lie in a figure F (including its boundary, of course) of area S. By Tool 8.2.7, there is a line through v_1 dividing F into two pieces, F_1 and F_2, of equal area $|F|/2$. By the Pigeonhole Principle, one of these pieces contains at least three of the five points $v_2, v_3, \ldots, v_6$. Without loss of generality, we can assume that F_1 contains the points v_2, v_3, and v_4 (note that both F_1 and F_2 contain the point v_1 on their boundary).

Now we can throw away the piece F_2 and apply the above reasoning to F_1. By Tool 8.2.7, there is a line through v_1 dividing F_1 into two pieces, F_{11} and F_{12}, of area $|F|/4$. By the Pigeonhole Principle, one of these pieces contains at least two of the three points v_2, v_3, and v_4. Without loss of generality, we can assume that F_{11} contains the points v_2 and v_3 (see Figure 8.10).

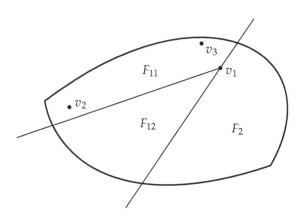

Figure 8.10

We are done; the area of the triangle $v_1 v_2 v_3$ does not exceed the area of F_{11}, which is equal to $|F|/4$.   □

Can we improve the inequality of Problem 8.2.11? Not really; there exists a figure F such that $S(F) = 6$. Can you find one such figure? Try it. And then compare your find with mine in Section 8.6.

Please note, we proved more than we stated in Problem 8.2.11. Here is what we actually proved:

Problem 8.2.12. Given six points $v_1, v_2, \ldots, v_6$ in a convex figure F, then every given point v_i is a vertex of a triangle $v_i v_j v_k$ of area at most $|F|/4$ formed by v_i and two other given points v_j and v_k.

This last result did not follow from Gideon's proof. In fact, I noticed it only now when I was writing these lines for you. It raises the question of whether something similar is true for five points in a triangle (see Problem 7.2.1).

Problem 8.2.13. Given five points $v_1, v_2, \ldots, v_5$ in a triangle of area 1, then is it true that *every* given point v_i is a vertex of a triangle $v_i v_j v_k$ of area at most $1/4$ formed by v_i and two more given points v_j and v_k?

The answer is no. Can you prove it, i.e., construct a counterexample?

We did the absolute best possible with the upper bound of $S(F)$. Now it is time to look for lower bounds.

8.3   Study of Convex Figures: A Lower Bound of $S(F)$

Obviously $S(F) \geq 3$: you need three points to form a triangle! But can $S(F)$ be equal to 3? No! Problem 8.3.1 will tell you why.

Problem 8.3.1. For any convex figure F,

$$S(F) \geq 4.$$

Solution. All we need to prove is that $S(F) \neq 3$, i.e., there exist three points M, N, P in the figure F such that

$$|MNP| > \frac{1}{4} |F|. \tag{46}$$

By Tool 8.2.7, there is a line L splitting F into two pieces of equal area. Let M and N be the points of intersection of L with the boundary of F. Draw supporting lines for F (Figure 8.11): one at M (which exists by Tool 8.2.8) and two parallel to MN (which exist by Tool 8.2.9).

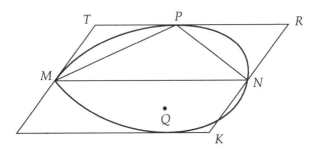

Figure 8.11

Let us also draw a line NK through N parallel to the supporting line at M. Since the figure F is convex, NK may not pass through an inside point of F both above and below MN. Without loss of generality, let us assume that NK does not pass through an inside point of F above MN. Denote the appropriate intersections by T and R (Figure 8.11).

We are done! If P is a point where the supporting line parallel to MN above MN touches the border of F, then we have found the required three points: M, N, P. Indeed, since exactly half the area of the figure F is completely inscribed in the parallelogram $MTRN$ (i.e., $|MTRN| \geq \frac{1}{2}|F|$) we get

$$|MNP| = \frac{1}{2}|MTRN| \geq \frac{1}{4}|F|. \qquad (47)$$

Are we really done? We needed to prove the strict inequality (46) but got only $|MNP| \geq \frac{1}{4}|F|$. Well, equality in (47) takes place only when the part of the figure F above MN coincides with the parallelogram $MTRN$. In this case, we can pick the points T, R and any point Q of the figure F below MN (Figure 8.11). You can easily prove (do) that $|TRQ| > \frac{1}{4}|F|$. □

We know now (Problems 8.2.11 and 8.3.1) that for any convex figure

$$4 \leq S(F) \leq 6,$$

i.e.,

$S(F)$ *can take on at most three values:* 4, 5, *or* 6!

Problem 8.3.1 serves as a starting point for two exciting new directions in our research. One path will take us to the best lower bound of $S(F)$ in Section 8.5; the other path will lead us to some exciting solved and unsolved problems in Section 8.7. First, however, I need to give you a very brief introduction to affine geometry.

8.4 A Very Brief Introduction to Affine Geometry

What is common between *translations, rotations,* and *reflections* of the plane? They all are *one-to-one transformations of the plane:* they move (map) distinct points of the plane into distinct points. They also preserve the size and shape of geometric figures. For that, all these transformations and their compositions (i.e., consecutive applications) are called *isometries*. An isometry maps a geometric figure F into a figure $f(F)$ congruent to F (see Figure 8.12).

Let me spoil this by adding homotheties.[2] As you may know, a *homothety f* with center of homothety O and coefficient of homothety k is the transformation that maps every point P into the point $P_1 = f(P)$ on the line OP such that

---

[2] You can find a few entertaining problems on homotheties in my book [S1] and an in-depth discussion of the subject in a marvelous book [Y4] by Isaac M. Yaglom.

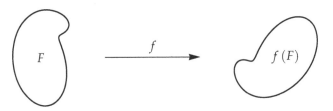

Figure 8.12

$$\frac{\left|\overrightarrow{OP_1}\right|}{\left|\overrightarrow{OP}\right|} = k.$$

Please note that unlike the case with the length of a segment, the direction of vectors $\overrightarrow{OP_1}$ and $\overrightarrow{OP}$ is taken into account; the measure $|\overrightarrow{a}|$ of a vector $\overrightarrow{a}$ is equal to the length $|a|$ of the corresponding segment a if the direction of $\overrightarrow{a}$ is the same as that of a unit vector, and is equal to $-|a|$ otherwise. A negative coefficient of homothety k would imply that the points P_1 and P are on opposite sides of the center of homothety O. Of course, the homothetic image $F_1 = f(F)$ of a geometric figure F under homothety f is the geometric figure consisting of homothetic images $P_1 = f(P)$ of all points P that make up the given figure F (see Figure 8.13).

Please note (and prove!) the following three properties of homotheties:

Problem 8.4.1. The homothetic image of a line is a line.

Problem 8.4.2. The homothetic image $f(P)$ of a polygon (or any geometric figure) P is similar to P.

Problem 8.4.3. Homotheties preserve the ratios of areas.

What properties of geometric figures are preserved under isometries, homotheties, and their compositions? Only the shapes of geometric figures; we lost the preservation of sizes.

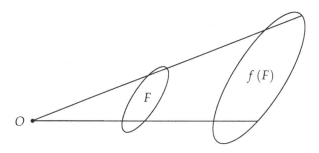

Figure 8.13

Now we give up even preserving shapes, and require only that lines be mapped into lines. What we get is the class of affine transformations.

Definition 8.4.4. An *affine transformation* is a one-to-one transformation of a plane into itself that maps lines into lines.

Isometries and homotheties are examples of affine transformations. So are parallel projections.

Definition 8.4.5. Given a plane P and a line L not parallel to P. A *parallel projection* f of P onto itself occurs when we move P in space to a new position P' (not parallel to L) and then project P' onto P parallel to L. I.e., every point p of P' is mapped into the point $f(p)$ of P such that the line $pf(p)$ is parallel to L (Figure 8.14).

Now I would like *you* to prove the following statements describing properties of parallel projections.

Problem 8.4.6. Parallel projections are affine transformations.

Problem 8.4.7. Parallel projections map parallel lines into parallel lines.

Problem 8.4.8. Parallel projections preserve the ratios of lengths of collinear segments.

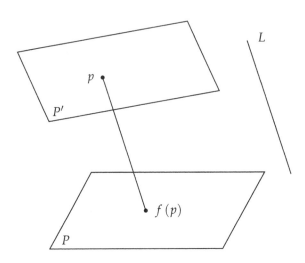

Figure 8.14

Problem 8.4.9. Parallel projections preserve the ratios of areas.

Problem 8.4.10. (I. Yaglom [Y4], Vol. 3, p. 13 of the English translation) Let A, B, C be three non-collinear points in a plane P, and let M, N, K be three non-collinear points in a plane P'. Then the planes P and P' can be placed in space so that there exists a parallel projection of P onto P' that maps the triangle ABC into a triangle $A'B'C'$ similar to the triangle MNK.

The following result is very important, but its proof is lengthy and not particularly beautiful; thus, I am presenting it to you here without proof.

Tool 8.4.11. (I. Yaglom, [Y4], Vol. 3, p. 18 of the English translation) There exists a unique affine transformation of the plane that takes three given non-collinear points A, B, C into three given non-collinear points A', B', C'.

Tool 8.4.11 and Problem 8.4.10 have the following great corollary (prove it!):

Problem 8.4.12. Every affine transformation can be obtained as a composition of a parallel projection followed by an isometry followed by a homothety.

Problems 8.4.12, 8.4.9, 8.4.8, and 8.4.3 have the following fundamental corollaries. Can you prove them?

Tool 8.4.13. Affine transformations preserve the ratios of lengths of collinear segments.

Tool 8.4.14. Affine transformations preserve the ratios of areas.

Tool 8.4.15. Affine transformations map parallelograms into parallelograms. Squares and parallelograms are indistinguishable in affine geometry: for any parallelogram P and square Q, there exists an affine transformation f such that $f(P) = Q$. Of course, in this case $f^{-1}(Q) = P$.

Tool 8.4.16. Affine transformations map ellipses into ellipses. Circles and ellipses are indistinguishable in affine geometry: for any ellipse E and circle C, there exists an affine transformation f such that $f(E) = C$. Of course, in this case $f^{-1}(C) = E$.

These are very powerful and convenient tools. Lots of exciting problems can now be solved easily and beautifully with their aid.

Problem 8.4.17. Prove that the medians of any triangle intersect at a point.

Solution. By an appropriate affine transformation f (Tool 8.4.11), we can map a given triangle T into an equilateral triangle T_0. Due to Tool 8.4.13, the midpoints of the sides of T will be mapped into the midpoints of the sides of T_0; therefore, the medians will be mapped into the medians. But the medians of the equilateral triangle T_0 intersect at a point, say O; therefore (since f is a one-to-one map!), the medians of T intersect at a point, namely $f^{-1}(O)$.   □

Problem 8.4.18. Given an ellipse E. Find a triangle T of maximum area inscribed in E.

Solution. There is an affine transformation f that maps E into a circle C (Tool 8.4.16). Of all triangles inscribed in C, the equilateral triangle T_0 has the maximum area (prove it!). Let $T = f^{-1}(T_0)$. Then T is a triangle inscribed in the given ellipse E.

   T is the required triangle, because according to Tool 8.4.14, the transformation f preserves the ratio of areas, i.e.,

$$\frac{|T|}{|E|} = \frac{|T_0|}{|C|}.$$

$\square$

   Please note: we can actually construct T by finding a cylinder with section E (can you think of a way to do that?) and defining f to be the orthogonal projection of E onto the base C of the cylinder. Can you finish this construction? I am sure you can!

Problem 8.4.19. Over all parallelograms of area 1, find the maximum area of an ellipse inscribed in a parallelogram.

Solution. First, let us prove that this maximum exists. Let P be a parallelogram of area 1. By Tool 8.2.5, there exists an ellipse E of maximum area inscribed in P.

   Since every parallelogram Q of area 1 can be mapped into P by an affine transformation f (Tool 8.4.15) and f preserves the ratio of areas (Tool 8.4.14), the maximum areas of ellipses inscribed in Q and P are equal. Therefore, the maximum area of an ellipse inscribed in a parallelogram exists over all parallelograms of area 1!

   Now let us find this maximum. There is an affine transformation f that maps E into a circle C (Tool 8.4.16). f maps the parallelogram P into the parallelogram $P_1 = f(P)$ circumscribed about the circle C. Now we can use the ratio of areas that is preserved under the transformation f (Tool 8.4.14):

$$\frac{|E|}{|P|} = \frac{|C|}{|P_1|}.$$

Thus, our problem is reduced to finding the maximum of $\dfrac{|C|}{|P_1|}$ over all parallelograms circumscribed about the circle C (Figure 8.15), which is attained (prove it!) when P_1 is a circumscribed square.

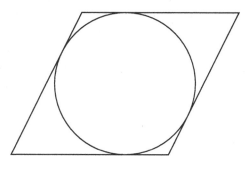

Figure 8.15

Simple computations (do them!) give us the maximum area $|E|$:

$$|E| = \frac{\pi}{4}.$$

□

And now, of course, a few problems for you, my reader, to enjoy on your own[3]:

Problem 8.4.20. Prove that the line through the intersection of the extensions of the non-parallel sides of a trapezoid and the intersection of its diagonals bisects both parallel bases of the trapezoid.

Problem 8.4.21. Given an ellipse E of area 1 and a positive integer $n \geq 3$. Find an n-gon N of maximum area inscribed in E. What is the area of N?

Problem 8.4.22. (G. D. Chakerian and L. H. Lange [CL]) In a given triangle T, inscribe an ellipse of maximum area.

---

[3] You will find more problems in I. Yaglom's [Y4], Vol. 3 of the English translation.

8.5 Study of Convex Figures: The Best Lower Bound of $S(F)$

There is a fundamental difference between mathematical problems in school (or college!) and problems mathematicians work on. A typical school problem sounds like this: "Given A, prove B" (sometimes even "given A, prove B using theorem C"!). Nobody gives "B" to a research mathematician; "B" is simply not known!

This is where you need intuition. Do you know what that is? Intuition is the ability to see the result (the "B"!) before you have the foggiest idea how to prove it! Intuition is extremely important in the work of a mathematician! You have to envision and conjecture the result, then cut in the steps and climb up to it. Imagine what would happen if you conjecture a false result; it would take a very long time to prove it, a very long time indeed!

As you recall, our results of Sections 8.2 and 8.3 limit possible values of the function $S(F)$ to no more than three: 4, 5, 6. Already in April 1988, when I had just obtained these results, I felt that, in fact, 4 may not be the value of $S(F)$ for any convex figure F.

Conjecture 8.5.1. $S(F) \neq 4$ for any convex figure F.

I sent this conjecture to a few young and not-so-young mathematicians and included it in my presentation "Beyond Problem Solving" at the Sixth International Congress on Mathematical Education in 1988 in Budapest. The conjecture proved to be true. I received two responses. A solution by Royce Peng was based on Tools 8.2.4 and 8.2.7. Semion Slobodnik, to whom I gave the conjecture during a meeting in August 1988 in Moscow, phoned me the next day (!) and over the phone gave me the following new and beautiful proof.[4] The result itself is not new. In fact, a good number of mathematicians have published its proofs: C. Radziszewski in his 1952 paper [R1], Wilhelm Süss three years later in [Sü], and eight years later Curtis M. Fulton and Serman K. Stein in [FS].

---

[4] A year later Semion handed me a written version of the proof. The one I am presenting here is that version in which I corrected an error and simplified the computational part.

Problem 8.5.2. (C. Radziszewski) For any convex figure F of area 1, there exists an inscribed parallelogram of area at least $\frac{1}{2}$.

Proof by S. Slobodnik. Let P be a parallelogram of maximum area inscribed in the figure F (which exists by Tool 8.2.5). According to Tool 8.2.6, the vertices of P lie on the boundary of F. We are going to prove that $ABCD$ is a parallelogram of the required area, i.e., that

$$|P| \geq \frac{1}{2}. \tag{48}$$

Since there is an affine transformation f which maps the parallelogram P into a square (Tool 8.4.15) and preserves the ratio $\frac{|P|}{|F|}$ of areas (Tool 8.4.14), we can assume without loss of generality that P is a square!

Our figure F consists of five convex parts (see Figure 8.16): the square P and four areas A, B, C, and D outside of P, each bounded by one side of P and the boundary of F.

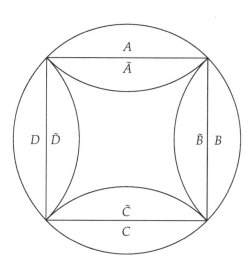

Figure 8.16

Now we fold in the parts $A, B, C,$ and D: let $\tilde{A}, \tilde{B}, \tilde{C},$ and $\tilde{D}$ be the symmetric images of areas $A, B, C,$ and D with respect to the sides of the square P bordering them. We want to prove that the figures $\tilde{A}, \tilde{B}, \tilde{C},$ and $\tilde{D}$ are *mutually area-disjoint*, i.e., that the intersection of any two of them has zero area. That would imply inequality (48) in the following way:

$$1 = |F| = |P| + (|A| + |B| + |C| + |D|)$$
$$= |P| + (|\tilde{A}| + |\tilde{B}| + |\tilde{C}| + |\tilde{D}|) \leq 2\,|P|$$
$$\text{i.e.,} \quad |P| \geq \frac{1}{2}.$$

We will consider two cases:

Case 1: Let the intersection N of two adjacent parts, say $\tilde{A}$ and $\tilde{D}$, have non-zero area. Then N contains three non-collinear points, so N contains a little circle O (see Figure 8.17).

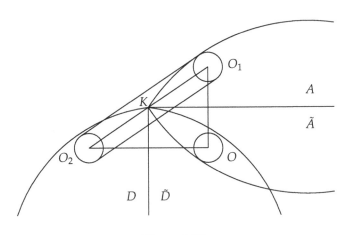

Figure 8.17

Denote by O_1 and O_2 the symmetric images of the circle O with respect to the sides of the square P bordering $\tilde{A}$ and $\tilde{D}$, respectively.

Since O_1 is completely in A and O_2 is completely in D, both O_1 and O_2 lie completely in the given figure F. Since F is convex, it must contain completely the *convex hull H* of the circles O_1 and O_2 (i.e., the minimal convex figure H that contains both O_1 and O_2).

It is easy to show (do!) that a vertex K of the square P is an inside point of H (i.e., K is included in H with a little circle around K), and, therefore, K is an inside point of F in contradiction to the fact that all vertices of P lie on the boundary of the figure F.

Case 2: Let the intersection N of two opposite parts, say $\tilde{A}$ and $\tilde{C}$, have non-zero area. Then N contains a little circle O (see Figure 8.18).

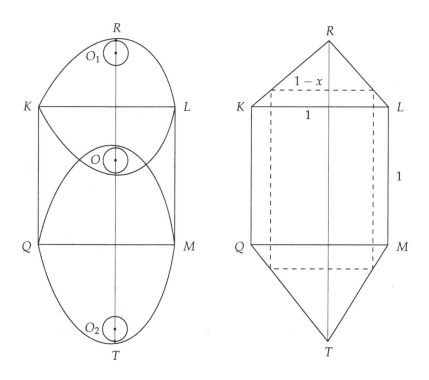

Figure 8.18   Figure 8.19

The symmetric images O_1 and O_2 of the circle O with respect to the upper and lower sides of the square P lie completely inside the given figure F. Denote the uppermost point of the circle O_1 by R and the lowermost point of the circle O_2 by T.

Since the square P is a parallelogram of maximum area inscribed in the figure F, P must be a parallelogram of maximum area inscribed in the hexagon $KRLMTQ$. We are going to find in this hexagon an inscribed rectangle of area greater than the area of P.

Since a homothety can make the sides of square P equal to 1, and homothety preserves the ratio of areas, we can assume without loss of generality that the sides of P are equal to 1. This will simplify our computations.

Now let us inscribe in our hexagon a rectangle Z with sides parallel to the sides of square P and horizontal side equal to $1 - x$ (see Figure 8.19). It is easy to notice that $|RT| = 2 + d$, where d is the diameter of circle O.

Using similar triangles, you can compute (do!) the other side of the rectangle Z: it is equal to $1 + (1 + d)x$. We are ready to compute the area of Z:

$$|Z| = (1 - x)\,[1 + (1 + d)x] = 1 + x\,[d - (1 + d)x]\,.$$

By choosing x small enough to make the expression in brackets positive, we can produce a rectangle Z of area greater than 1, i.e., greater than $|P|$, in contradiction to the maximality of area of P.

We are done. $\square$

This is a very nice proof. But, there is a "but." I asked Semion to prove Conjecture 8.5.1. The inequality $S(F) \neq 4$ means that in any convex figure F of area 1, there are four points such that any three of them form a triangle of area *greater* than $\frac{1}{4}$. Problem 8.5.2 guarantees the existence of four points (vertices of an inscribed parallelogram of maximum area) such that any three form a triangle of area *greater than or equal to* $\frac{1}{4}$.

What do we do?

No problem! I mean, we can solve the problem:

Problem 8.5.3. For any convex figure F,

$$S(F) \neq 4.$$

Solution. Slobodnik's solution of Problem 8.5.2 leaves open only the case when $|P| = \dfrac{1}{2}$, i.e., when parts $\tilde{A}$, $\tilde{B}$, $\tilde{C}$, and $\tilde{D}$ *completely cover the square P* (see Figure 8.16). Since all parts $\tilde{A}$, $\tilde{B}$, $\tilde{C}$, and $\tilde{D}$ are *convex*, the borders between them must be line segments! In fact, no more than one segment can serve as a border between any two parts $\tilde{A}$, $\tilde{B}$, $\tilde{C}$, and $\tilde{D}$!

We will consider two cases.

Case 1: Let a pair of opposite parts, say $\tilde{A}$ and $\tilde{C}$, have a border segment $\overline{XY}$ (X to the left of Y) in common (see Figure 8.20).

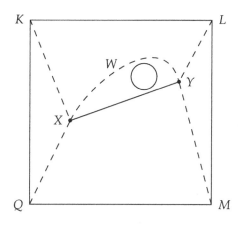

Figure 8.20

The case when X and Y coincide with two vertices of the square is very simple (do it). Assume now that X does not coincide with any of the vertices of the square. Without loss of generality, we can assume that Y may coincide with L but not with M. While we keep $\tilde{A}$ (i.e., the quadrilateral $KXYL$) unchanged, we enlarge $\tilde{C}$ (i.e., the quadrilateral $QXYM$) by replacing segment $\overline{XY}$ with arc XWY.

Now new parts $\tilde{A}$ and $\tilde{C}$ have intersection N of positive area; therefore, N contains a little circle, and we can word-for-word repeat Case 2 of Slobodnik's proof of Problem 8.5.2 up to the contradiction with the maximality of area of the square $KLMQ$.

Case 2: Now let us assume that opposite parts $\tilde{A}$ and $\tilde{C}$, $\tilde{B}$ and $\tilde{D}$ do not have more than one point in common (see Figures 8.21 and 8.22).

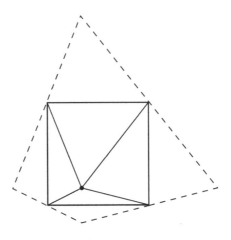

Figure 8.21

When we unfold $\tilde{A}$, $\tilde{B}$, $\tilde{C}$, and $\tilde{D}$, we will see (prove it!) that the original figure F is a triangle or a quadrilateral.

There is nothing we can do to enlarge the inscribed square in the case of a triangle F. Fortunately, we do not need to: Problem 7.1.5 delivers the necessary choice of four points.

If the original figure F is a quadrilateral, we draw its two diagonals and look at the areas of the four triangles formed by a diagonal and two sides of F, and consider three cases.

If the area of each is greater than $\dfrac{1}{4}|F|$, then the four vertices of F are the required set of four points (Figure 8.23).

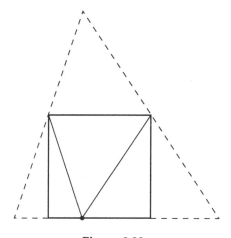

Figure 8.22

If the area of one of the four triangles is less than $\frac{1}{4}|F|$, we throw it away, and in the remaining triangle we pick the four points: its three vertices and the center of mass, as in Problem 7.1.5 (Figure 8.24). You can easily verify that any three of the selected four points form a triangle of area greater than $\frac{1}{4}|F|$.

If each diagonal of the quadrilateral F cuts off a triangle of area exactly $\frac{1}{4}|F|$, then F is a trapezoid with ratio of parallel bases $1:3$ (Figure 8.25). You can easily prove that using Tool 7.3.1 (Hint: $|BAC| = |BDC|$ implies $AD \parallel BC$).

Without loss of generality, we can assume that the parallel bases of F have $|BC| = 1$, $|AD| = 3$, and altitude $|CE| = 1$. Why? Because there is an affine transformation f which maps F into a trapezoid with required dimensions (prove it!), and f preserves the ratio of areas. We do this to simplify our computations.

Now let MN be parallel to BC and a "little" longer than BC (Figure 8.26). I.e., let $|MN| = 1 + x$. It is not hard to compute the altitude of the quadrilateral $MNDA$ (do!):

$$|FE| = 1 - \frac{x}{2}.$$

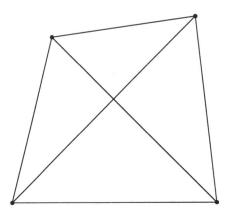

Figure 8.23

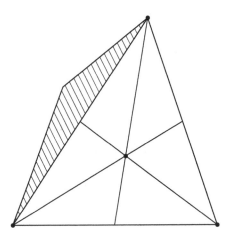

Figure 8.24

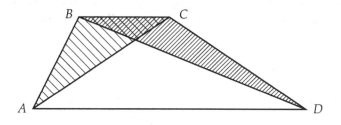

Figure 8.25

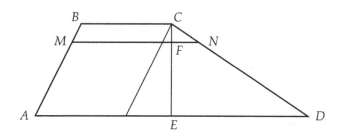

Figure 8.26

We are ready to compute the areas:

$$|F| = |ABCD| = 2$$

$$|MAN| = |MDN| = \frac{1}{2}(1+x)\left(1 - \frac{x}{2}\right)$$

$$= \frac{1}{2}\left(1 + \frac{x}{2} - \frac{x^2}{2}\right) = \frac{1}{4}|F|\left[1 + \frac{x}{2}(1 - x)\right].$$

You can see that if you choose x to be a small positive number, say $x = \frac{1}{100}$, then the areas of the triangles MAN and MDN will be slightly greater than $\frac{1}{4}|F|$, and thus we can pick the required set of four points: M, N, D, A.

We are done! $S(F) \neq 4$ for convex figures F!   □

In the process, we proved (can you see how?) another nice result:

Problem 8.5.4. For any convex figure F that is not a triangle or a quadrilateral, there exists an inscribed parallelogram of area greater than $|F|/2$.

A stronger result, however, was obtained in 1960:

Problem 8.5.5. (C. Fulton and S. Stein [FS]) If F is a convex figure, then there is a parallelogram inscribed in F of area greater than $|F|/2$ if and only if F is not a triangle.

I would like to point out here that Problems 8.5.5 and 7.1.5 imply the result of Problem 8.5.3. I chose to include a direct proof of Problem 8.5.3 because it demonstrates how Slobodnik's proof of Problem 8.5.2 can be strengthened to prove Conjecture 8.5.1.

8.6 A One-Hundred-Dollar Problem

In Sections 8.2 and 8.5, we proved that the function $S(F)$ on convex figures F can take on at most two values: 5 and 6. We had examples of figures F for which $S(F) = 5$ (Problems 7.2.1 and 8.1.2). But can $S(F)$ be equal to 6 for a convex figure F?

Problem 8.6.1. Is there a convex figure F such that

$$S(F) = 6?$$

Can you solve this problem? No? Well, it is a tough one! What did you try to prove, that "there is" or that "there isn't?" Let me help you: there is.

Problem 8.6.2. Find an example of a convex figure F such that

$$S(F) = 6.$$

Now, after you have tried to find an example, I would like to share with you mine:

Problem 8.6.3. Prove that for a regular pentagon F,

$$\mathbf{S}(F) = 6.$$

Solution. We have to prove that $\mathbf{S}(F) > 5$, i.e., that there is a set of five points such that each of the ten triangles formed by them has area greater than $\frac{1}{4}|F|$.

The five vertices of F deliver such a set. It is not difficult to compute (do!) the area S of the smallest of the ten triangles (Figure 8.27).

S comes out to be large enough:

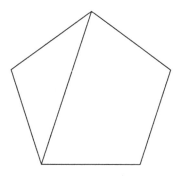

Figure 8.27

$$S = \frac{5 - \sqrt{5}}{10}|F| > \frac{1}{4}|F|.$$

What values does $\mathbf{S}(F)$ take on for some familiar figures F?

Problem 8.6.4. Find $\mathbf{S}(F)$ for a circle F.

Problem 8.6.5. Find $\mathbf{S}(F)$ for a regular hexagon F.

Our games in $\mathbf{S}(F)$ have really paid off. The function $\mathbf{S}(F)$ takes on exactly two values: 5 and 6. Accordingly, it splits the universe of convex figures into two classes! It would be great to classify these two classes, to have a clue as to why some convex figures F get the invariant value $S(F) = 5$ and others get 6. In July 1989, I offered a fifty-dollar prize for the first solution to this problem. Five years later, I raised the prize to one hundred dollars.[5]

The One-Hundred-Dollar Problem 8.6.6. Find (and classify) all convex figures F such that

$$S(F) = 6.$$

So far, nobody has claimed the prize.

You have a chance! Try it, but remember: it is an open problem; nobody has solved it yet, and it may be very difficult.

Here is what I expect to be the answer:

Conjecture 8.6.7. Let F be a convex figure. $S(F) = 6$ if and only if there is an affine transformation f that maps the boundary of F into a fairly narrow frame (width to be determined) made by two concentric regular pentagons (Figure 8.28).[6]

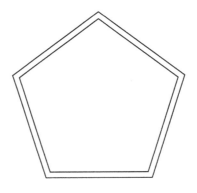

Figure 8.28

[6] In 2007, my student and coauthor Dmytro Karabash produced a counterexample to this conjecture—see the new Chapter 14 later in this book. Dmytro has not proposed an alternative conjecture.

The condition in my conjecture is clearly sufficient: if we alter a regular pentagon "very little," then the value of the function $\mathbf{S}(F)$ will remain 6.

The necessary part is much less obvious. Why should all convex figures F such that $\mathbf{S}(F) = 6$ have only one ancestor, a regular pentagon?

Well, it is *my* hypothesis! If you do not believe it, come up with your own.

Then prove it!

In my lecture at the Evanston Township High School, I presented Problem 8.6.6 in a short form:

$$\text{Solve the equation} \quad \mathbf{S}(x) = 6.$$

Saul Kato, a junior, immediately responded with a short answer:

$$x = \mathbf{S}^{-1}(6).$$

This is a short section. It is missing the main result of this chapter: a solution to Problem 8.6.6.

Hopefully, one of us will fill in the blank soon.

8.7 Triangle in Ellipse

I would like to bring your attention to a spin-off of our research. Problem 8.3.1 can be reformulated as follows:

Problem 8.7.1. Any convex figure F contains an inscribed triangle of area greater than $\frac{1}{4}|F|$.

Can we improve this lower bound $\frac{1}{4}|F|$? There are known results in this area, but before I give them to you, I would like to present here a result obtained by the Russian mathematician and my friend, Semion Slobodnik. His result is not as good as the known results, but his proof is so beautiful that I have to share it with you.

Semion and I first met when we were both 14. At that time, we had met every April and competed at the Moscow University Mathematical Olympiad. Later, when we both were undergraduates (and, therefore, could not compete in the Olympiad any more!), Semion challenged me to a duel. He did not throw a glove in my face; he simply told me "there is no better problem solver in the world than I, and that includes you." We proceeded with a Renaissance-style mathematical duel. The choice of weapons was mine: we gave each other one problem and 24 hours to solve it. The next day, I showed him a solution to his problem; he showed me that my problem was, in fact, a corollary to his!

I left Russia in 1978. "Ten years later" (as the author of *The Three Musketeers* Alexandre Dumas would have said), I visited Moscow and discovered that once again Semion and I were working independently on related problems! Semion gave me the following result with the inscription, "To Sasha from Sema." I hope you will enjoy it as much as I did.

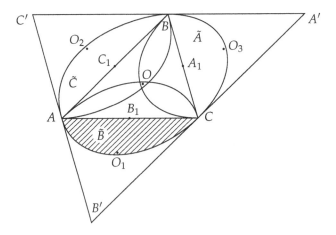

Figure 8.29

Problem 8.7.2. (S. Slobodnik) Any convex figure F contains an inscribed triangle of area at least $\dfrac{1}{3}|F|$.

Solution. Let ABC be a triangle of maximum area inscribed in the figure F (it exists due to Tool 8.2.5). Through the vertices A, B and C, we draw lines parallel to the opposite sides of the triangle (see Figure 8.29).

We get the triangle $A'B'C'$, which contains the figure F completely. Indeed, if we assume that the figure F rises above $A'C'$ (see Figure 8.30), then ADC will have area greater than that of ABC, in contradiction to the maximality of the area of ABC.

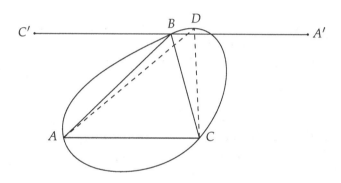

Figure 8.30

Let C_1, A_1, and B_1 be the midpoints of the sides $\overline{AB}$, $\overline{BC}$, and $\overline{CA}$, respectively (Figure 8.29). $\tilde{B}$ will denote that part of figure F contained in the triangle $AB'C$, and $\tilde{\tilde{B}}$ will be the symmetric image of $\tilde{B}$ with respect to B_1. The figure $\tilde{\tilde{B}}$ lies completely inside the triangle ABC because ABC is symmetric to the triangle $AB'C$ with respect to B_1. Similarly, we define figures $\tilde{C}$ and $\tilde{A}$ and their images $\tilde{\tilde{C}}$ and $\tilde{\tilde{A}}$ with respect to C_1 and A_1.

We want to prove that the intersection N of $\tilde{A}$, $\tilde{B}$, and $\tilde{C}$ has zero area:

$$|N| = 0. \tag{49}$$

Assume that this is not true, i.e., $|N| > 0$. Then N contains three points a, b, c not all on a line. Since N is the intersection of convex figures $\tilde{A}$, $\tilde{B}$, and $\tilde{C}$, N itself is convex and, therefore, N contains the triangle abc, which implies that N contains a little circle R with center O.

Let O_1, O_2, and O_3 be the symmetric images of O with respect to B_1, C_1, and A_1. Then the triangle $O_1O_2O_3$ lies inside F and is congruent to ABC—this is a fun plane geometry problem in itself. Try to finish it and compare your answer with my proof below (Problem 8.7.3).

If we now map the little circle R with center O symmetrically with respect to each of the points B_1, C_1, and A_1, we get three little circles (with centers O_1, O_2, and O_3) that lie completely inside F. Now we can find a triangle in F that contains $O_1O_2O_3$ and has area greater than the area of $O_1O_2O_3$ (Figure 8.31), i.e., greater than the area of ABC. This is in contradiction to the maximality of the area of ABC.

The equality (49) is proven. This means that every region of the triangle ABC belongs to at most two of the three figures $\tilde{A}$, $\tilde{B}$, and $\tilde{C}$. Now we are ready to calculate the areas.

$$|\tilde{A}| + |\tilde{B}| + |\tilde{C}| \leq 2\,|ABC|.$$

Therefore,

$$\begin{aligned} |F| &= |ABC| + |\tilde{A}| + |\tilde{B}| + |\tilde{C}| \\ &= |ABC| + |\tilde{A}| + |\tilde{B}| + |\tilde{C}| \leq 3\,|ABC|, \end{aligned}$$

which implies the desired inequality $|ABC| \geq \dfrac{1}{3}|F|$.   □

Now we have to pay our debt: let us prove a result that we have already used in Problem 8.7.2.

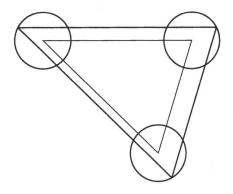

Figure 8.31

Problem 8.7.3. Given a triangle ABC and a point O inside it. Points O_1, O_2, O_3 are the symmetric images of O with respect to the mid-points A_1, B_1, C_1 of the sides of ABC (see Figure 8.32). Prove that the triangles $O_1O_2O_3$ and ABC are congruent.

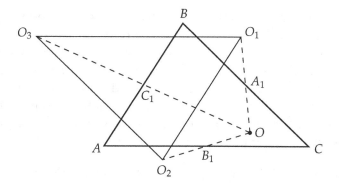

Figure 8.32

Solution. $O_1O_2O_3$ is the homothetic image of the triangle $A_1B_1C_1$ with coefficient of homothety 2 and center O. The triangle ABC is the homothetic image of the same triangle $A_1B_1C_1$ with coefficient of homothety -2 and center at the center of mass (the intersection of medians) of ABC. Therefore, $O_1O_2O_3$ and ABC are congruent.   □

As I mentioned before, there are better estimates. One of them belongs to the famous Russian geometer and author of numerous wonderful books, Isaac Moiseevich Yaglom. I am presenting it here without proof for three reasons: (1) it has already been published in two books by I.M. Yaglom; (2) the proof is not as striking (in my opinion) as the one above; and (3) I want to allow you the pleasure of coming up with your own proof!

Problem 8.7.4. ([Y2], problem 121 and [Y3], problem 63) Let F be a convex polygon of area S, and L an arbitrary line. Prove that:

   a) there exists a triangle inscribed in F such that its area is at least $\frac{3}{8}S$ and one of its sides is parallel to L;

   b) F and L can be chosen in such a way that there is no triangle inscribed in F such that its area is greater than $\frac{3}{8}S$ and one of its sides is parallel to L.

You can certainly replace the words "convex polygon" in Yaglom's result with "convex figure" (Professor Yaglom probably did not want to over-complicate his book with things like Tool 8.2.5).

The result of Problem 8.7.4 is optimal in the setting that includes line L. Without line L, the best lower bound for the maximum area of an inscribed triangle was obtained by Wilhelm Blaschke over 70 years ago. I am grateful to Professor Branko Grünbaum for providing me with a copy of this 1917 paper.

Problem 8.7.5. (W. Blaschke, [B1]) Let T be a triangle of maximum area inscribed in a convex figure F. Then

$$|T| \geq \frac{3\sqrt{3}}{4\pi}|F|.$$

Moreover, the equality is attained if and only if F is an ellipse.

Do you know where $\dfrac{3\sqrt{3}}{4\pi}$ came from? It is the ratio of areas of an equilateral triangle T and the circle circumscribed about T (as you remember from Problem 8.4.16, circles and ellipses are indistinguishable in affine geometry).

Problem 8.7.5 shows that the worst convex figure to be approximated by a triangle is an ellipse. This consideration inspired two famous geometers, L. Fejes Tóth and Branko Grünbaum, to come up with conjectures. Accordingly, our research splits in two directions.

The famous Hungarian geometer, L. Fejes Tóth, conjectured (probably in the 1930s) that an ellipse is the worst figure to be approximated, not only by a triangle, but by an n-gon for any integer n, $n \geq 3$. And I am sorry to inform you that you were born too late to be the first to prove this conjecture: the Hungarian mathematician Sas published the proof in his 1941 paper [Sa]. It is amazing that Paul Erdős remembered this article, which he read almost fifty years ago, and its author, who did not survive World War II. The paper appeared in Hungarian only. But cheer up: L. Fejes Tóth included Sas' proof in his 1953 book [FT] which appeared in German and later was translated into Russian.

Problem 8.7.6. (Sas [Sa]) Let P be an n-gon of maximum area inscribed in a convex figure F (n is an integer, $n \geq 3$). Then

$$|P| \geq \frac{n}{2\pi} \sin \frac{2\pi}{n} |F|.$$

Moreover, the equality is attained if and only if F is an ellipse.

Can you guess where $\dfrac{n}{2\pi} \sin \dfrac{2\pi}{n}$ came from? You are right: it is the ratio of areas of a regular n-gon P and the circle circumscribed about P.

Professor Branko Grünbaum from the University of Washington conjectured in his letter to me that this pair (triangle–ellipse) provides the worst possible approximation of one convex figure by another.

Definition 8.7.7. (B. Grünbaum, personal communication) Let F, G be convex figures. We define the symbol $f(F, G)$ to be the maximum value of $\dfrac{|F^*|}{|G|}$, where F^* is an affine image of F such that F^* is contained in G.

In this notation, Problems 8.7.6 and 8.5.2 can be rewritten as follows:

Problem 8.7.8. Let Δ be a triangle. Then over all convex figures G,

$$\min f(\Delta, G) = \frac{3\sqrt{3}}{4\pi}.$$

The equality is attained only when G is an ellipse.

Problem 8.7.9. Let P be a parallelogram. Then over all convex figures G,

$$\min f(P, G) = \frac{1}{2}.$$

The equality is attained only when G is a triangle.

Open Problem 8.7.10. (B. Grünbaum, personal communication) What is the absolute minimum of $f(F, G)$ (i.e., the minimum over all convex figures F and G), and for what F and G is it attained?

Conjecture 8.7.11. (B. Grünbaum, personal communication) The absolute minimum of $f(F, G)$ is $\dfrac{3\sqrt{3}}{4\pi}$. It is attained only when F is a triangle and G is an ellipse.

Grünbaum's problem is very exciting, don't you think? I am offering a Twenty-Five-Dollar Prize for the first solution to it! (Of course, there are easier ways to earn money, but they are hardly as glamorous.)

9

Paul Erdős: Our Joint Problems

9.1 PGOM Erdős

Then came Paul Erdős. Have you heard of him and his famous problems and prizes?

A great big book could be written about this unique human being. His list of over 1500 publications would fill up many books! And his list of honorary degrees and memberships in academies. And his compassionate stand on human rights. And his love for children, whom he tenderly called epsilons. And his constant travel throughout the world to engage and inspire mathematicians young and old wherever he went!

Paul Erdős was a marvelous mathematician, author of numerous results that so expressively demonstrate the beauty, surprises and elegance of mathematical thought, results that "come straight from the Book.[1]" But it is his problems that won him a unique place in the history of mathematics.

Paul Erdős was the greatest problem creator of all times. Myriads of his problems have inspired and continue to inspire myriads of mathematicians. What is the secret? The Erdős problems are, just like classical problems of mathematics, short and clear, usually understandable by everyone, including young high school mathemati-

---

[1] See the epigraph on page xxvii of this book.

A. Soifer, *How Does One Cut a Triangle?*, Second Edition, DOI 10.1007/978-0-387-74652-4_9,   107
© Alexander Soifer 2009

Paul Erdős (left) and Alexander Soifer, Colorado Springs,
March 18, 1989. Photograph by Tom Kimmel.

cians. You take on a charming Erdős problem, and very soon you
can find some insight and prove some partial advances. Yet many
of the Erdős problems have not been solved for decades, and some
may withstand centuries; time will tell.

Where do you find Erdős problems? That is a problem, too. You
have to dig into research journals and the proceedings of confer-
ences (a small list of them is in the bibliography at the end of this
book). This is why I worked very hard to convince Paul to write a
book, *Problems of PGOM Erdős*, a book of unsolved problems created
by Erdős with his commentaries, a book for mathematicians young
and old to enjoy, a book to lead us and future mathematicians in our
research.[2]

---

[2] [Added in second ed.] On the advice of Ronald L. Graham, Paul Erdős invited me to join him in writing this book. It was in the works until his passing
in 1996, and I will finish our joint book as soon as I can, now that *The Mathematical Coloring Book* has been published [S5] .

Now, you want to know what "PGOM" is, right? In August 1988, when I visited Paul in his office at the Institute of Mathematics in Budapest, he presented me with a few of his reprints, on which he wrote "P. Erdős, PGOM." To my natural question, Paul explained, "poor great old man." "Great"—no questions about it, "old"—in age only—he was 75 years of age then, but why "poor?" I asked. "All old men are poor," answered Paul. One thing is clear, mathematicians are rich because of Paul Erdős, PGOM!

9.2 Problems

In March of 1989, Paul Erdős visited me here in Colorado Springs. We discussed problems you found on the preceding pages of this book. Paul was especially interested in the Five-Point Problem (Problem 7.2.1). Our visit produced three joint Twenty-Dollar Problems. Paul included one of these problems in his talk the following day at the University of Colorado at Colorado Springs. He offered $10 for the solution during the talk, and I, as a coauthor, immediately offered to match it.

Each of the problems asks for a "maximin." Each deals with a similar situation when a set S of five points is present. Since every three-point subset of S forms a triangle, we get $\binom{5}{3} = 10$ distinct triangles.

Let $\min\Delta(S)$ denote the minimum area of a triangle among the ten triangles determined by the set S.

Problem 9.2.1. (Twenty-Dollar Problem ES1) Over all triangles T of area 1 and all 5-point subsets S of T, find

$$L_T = \max\left(\min\Delta(S)\right).$$

Please note, Problem 7.2.1 shows that

$$\max\left(\min\Delta(S)\right) \leq \frac{1}{4}.$$

Problem 9.2.2. (Twenty-Dollar Problem ES2) Over all convex quadrilaterals Q of area 1 and 5-point subsets S of Q such that S consists of the four vertices of Q plus another point, find

$$L_Q = \max\left(\min \Delta\left(S\right)\right).$$

Problem 9.2.3. (Twenty-Dollar Problem ES3) Over all convex pentagons P of area 1, find

$$L_P = \max\left(\min \Delta\left(S\right)\right),$$

where S is the set of the vertices of P.

Here Paul Erdős and I both felt that the regular pentagon delivers the optimum solution:

Conjecture 9.2.4. (ES) $L_P = \dfrac{5 - \sqrt{5}}{10}$

Since that time, I have offered these three open problems (and prizes!) to many young and not-so-young mathematicians from several countries. A number of them attempted to solve them. I have received three replies so far, which we will discuss in the next section.

9.3 Solutions

And the youngsters are winning!

My summer 1989 student, Vladimir Baranovski, a senior high school student from Omsk, a city in Russian Siberia, and a candidate for the Soviet Union team for the International Mathematical Olympiad, sent me his solutions to ES2 and ES3 in a 12-page letter, which he mailed in September 1989 and I received over a month later (very slow airplanes, or slow censors!).

Royce Peng, a senior high school student from California, USA, whom you already met on the pages of this book, sent me a solution to ES1 in a 10-page letter in November 1989.

In December 1989, I received a letter from my summer 1989 student, Boris Dubrov, a senior high school student from Minsk, USSR, and the winner of the 1989 Soviet Union Mathematical Olympiad. He sent me the answers and optimal constructions for all three problems and promised to send his solutions soon.

It is interesting that these three talented young mathematicians might meet at the International Mathematical Olympiad in Beijing, China in July 1990 representing the USSR and the USA![3]

Their solutions are long and include a number of sub-cases, computations, and analytical geometry. This is why I am not including their complete proofs. I am going to present outlines of their solutions. I hope you will find and share with me alternative solutions to these problems.

Tool 9.3.1. Given an angle α, $0 < m(\alpha) < \pi$, and a point O inside it. Find the line through O that cuts out of angle α the triangle of minimum area.

Answer: The line L through O such that O is the midpoint of the segment cut out of L by the sides of the angle α (proof, see [S1], Problem 4.6.2).

Problem 9.3.2. (ES1, R. Peng)

$$L_T = \max\left(\min \Delta\left(S\right)\right) = \left(\sqrt{2} - 1\right)^2,$$

i.e., among any five points in any triangle T of area 1, there are three points that form a triangle of area at most $(\sqrt{2} - 1)^2$.

Moreover, the equality is attained if and only if there is an affine transformation that maps T and the five points into configurations described in Figures 9.4 and 9.5.

---

[3] [Added in second ed.] Indeed, Volodya Baranovski and Royce Peng met at the 1990 IMO! Vladimir Baranovsky is now an assistant professor of mathematics at the University of California, Irvine. Boris Dubrov is now a senior researcher at the N.G. Chebotarev Research Institute of Mathematics and Mechanics of the Kazan State University, Russia. A few years ago, when I last heard from Royce Peng, he was finishing his Ph.D. thesis in mathematics at the University of Southern California.

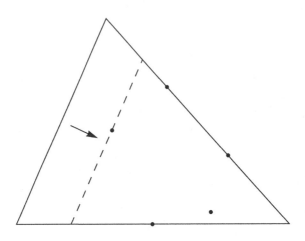

Figure 9.1

Outline of Royce Peng's Solution. Since we are trying to maximize the ratio of the areas of the smallest triangle and the triangle T that contains all five points, we should minimize the area of T. This implies that each side of the triangle T must pass through at least one of the given five points; otherwise, we can move a side parallel to itself until it hits a given point, and thus reduce the area of T (Figure 9.1).

Moreover, if a side AB of T passes through just one given point G, and G is not the midpoint of AB, then we can rotate AB in an appropriate direction, which will reduce the area of T (see Tool 9.3.1 and similar reasoning) until AB passes through one more given point H (Figure 9.2) or G becomes the midpoint of AB (Figure 9.3).

Thus, each side of T contains a given point in its middle or passes through two given points. That creates seven cases, which Royce considers.

He uses an affine transformation to transform the triangle T into an equilateral triangle and other clever simplifications. For example, he considers symmetric configurations of five given points first and then reduces general configurations to symmetric ones.

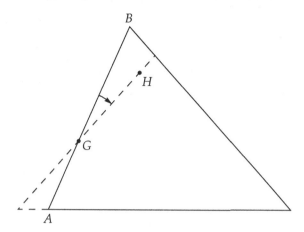

Figure 9.2

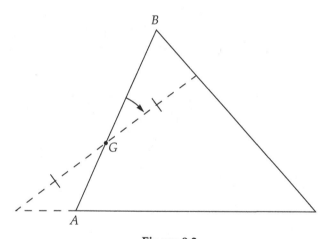

Figure 9.3

Royce classifies configurations for which the minimum area of a triangle is indeed equal to $(\sqrt{2} - 1)^2$. He obtains one example (Figure 9.4) and an infinite series of examples (Figure 9.5). In both, T is an equilateral triangle with side 1. Other sizes are marked on Figures 9.4 and 9.5.   □

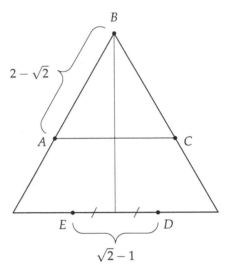

Figure 9.4 The five points are symmetric with respect to the altitude. The minimum area triangles are $|BCD| = |CDE| = |DEA| = |EAB| = (\sqrt{2} - 1)^2$.

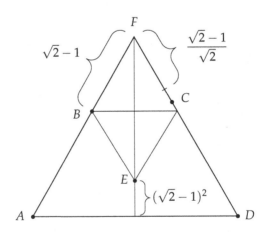

Figure 9.5 A, B, D, E have fixed positions; C can be anywhere, so that $\dfrac{\sqrt{2} - 1}{\sqrt{2}} \leq |FC| \leq \sqrt{2} - 1$. The minimum area triangles are $|AED| = |BEC| = |BED| = (\sqrt{2} - 1)^2$.

On the way to solving Problems ES2 and ES3, Vladimir Baranovski implemented a multiple use of Tool 7.3.1 to build a beautiful tool which he called a "Useful Lemma."

Tool 9.3.3. ("Useful Lemma" by V. Baranovski) Given a set S of five points. Prove that the points can be moved around in such a way that

1) the minimum area $\min \Delta\,(S)$ of a triangle with vertices in the points of S will not change;
2) the area of the convex hull of S (i.e., the minimal convex figure that contains S) will not increase;
3) in a new position, every point of the set S will serve as a vertex for at least two triangles of area $\min \Delta\,(S)$.

This is a very interesting tool. In true Mathematics Olympiad style, Volodya uses it to show in the settings of ES2 and ES3 that there is an optimal configuration with at least four triangles of minimum area, and that two of them have an edge in common (try to prove it). This allows him to very cleverly reduce all possible configurations to just a few, which he studies with the use of analytical geometry.

The optimal quadrilaterals Volodya obtains to demonstrate $L_Q = \frac{1}{2}\left(\sqrt{2}-1\right)$ are in Figures 9.6 and 9.7.

Problem 9.3.4. (ES2, V. Baranovski)

$$L_Q = \max\left(\min \Delta\,(S)\right) = \frac{1}{2}\left(\sqrt{2}-1\right)$$

"Useful Lemma" and analytical geometry allow Volodya Baranovski to solve Problem ES3 as well. His optimal pentagon for ES3 can be transformed by an affine transformation into a regular pentagon.

Volodia left a hole in his proof of "Useful Lemma." But since I was able to patch it up, Paul Erdős and I decided to award him $40 for ES2 and ES3. The $20 prize for ES1 goes to Royce Peng.

"Slow airplanes" may have prevented Boris Dubrov's solutions from arriving in time to be included. His letter, which I received on

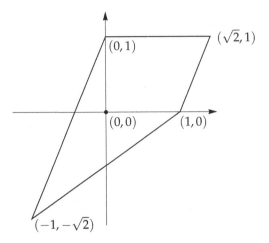

Figure 9.6

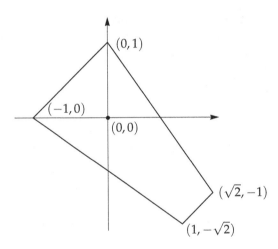

Figure 9.7

New Year's Eve, included correct answers to all three problems. It also contained a striking observation: the five points in Figure 9.8 deliver an optimal configuration for *both* problems ES1 and ES2!

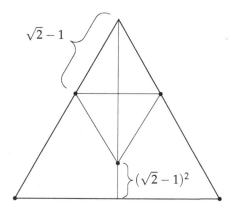

Figure 9.8

Those brilliant youngsters awakened my competitive spirit! I set out to find a nice, short, non-computational solution of ES3. I had to compete with time, also: the book was almost ready to go to print. I asked my secretary for one day. Here is what came out of that day:

Problem 9.3.5. (ES3)

$$L_P = \max \min \Delta\,(S) = \frac{1}{10}(5 - \sqrt{5}).$$

Moreover, the maximin is attained for a regular pentagon P, and only for it.

Solution. Given a convex pentagon P, its five vertices A, B, C, D, E determine $\binom{5}{3} = 10$ triangles. Let ABC be a triangle of the smallest area. Since for any pentagon there is an affine transformation that maps its triangle of smallest area into ABC, and since we are trying to maximize the ratio $|ABC| : |ABCDE|$, we can fix the points $A, B,$ and C and move the points D and E in such a way that

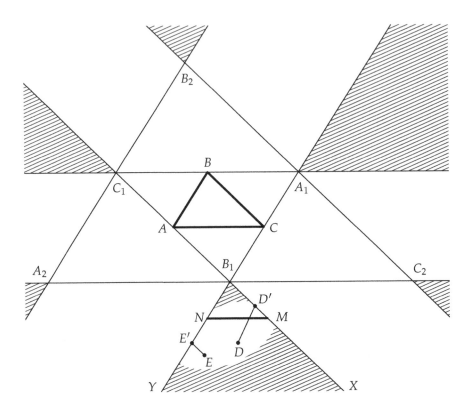

Figure 9.9

a) the area of the pentagon does not change;
b) ABC retains the smallest area among the 10 triangles formed by the vertices of the pentagon.

But first of all, let us explore the question: given ABC, where can D and E be located?

Through each vertex of ABC, we draw a line parallel to the opposite side of ABC. We get the triangle $A_1B_1C_1$ (see Figure 9.9). Now through each vertex of $A_1B_1C_1$ we draw a line parallel to the opposite side of $A_1B_1C_1$. We get the triangle $A_2B_2C_2$.

By Tool 7.3.1, the vertices D, E of the pentagon $ABCDE$ may not lie inside triangle $A_2B_2C_2$. They must be inside or on the boundaries of six striped angles. Moreover, due to convexity of the pentagon, D and E must both lie in the same angle A_1, B_1, or C_1. Let it be the angle XB_1Y. Assume for definiteness that the sequence of vertices of the pentagon is clockwise A, B, C, D, E.

Now we are ready for two moves, each of which satisfies the conditions (a) and (b) above:

1. move D parallel to EC to its new position D' on B_1X;
2. move E parallel to AD' to its new position E' on B_1Y.

Now we plot two points M and N on B_1X and B_1Y, respectively, so that $|MNC| = |MNA| = |ABC|$. Of course, MN is parallel to AC (Tool 7.3.1). What is much more exciting,

$$|ABC| = |BCM| = |CMN| = |MNA| = |NAB|,$$

and, therefore (prove it!), there is an affine transformation that maps $ABCMN$ into a regular pentagon!

All that is left to prove is

$$|ABCDE| > |ABCMN|, \tag{50}$$

unless, of course, $D = M$ and $E = N$.

Both points M, N may not lie outside the pentagon $ABCD'E'$ because that would imply $|E'CD'| < |NCM| = |ABC|$ in contradiction to (b) above.

If both points M, N lie inside the pentagon $ABCD'E'$, then $|ABCDE| = |ABCD'E'| > |ABCMN|$.

Let exactly one of the points M, N, say N, lie inside the pentagon $ABCD'E'$ (see Figure 9.10). Then

$$|ACD'E'| = |ACE'| + |CD'E'| > |CAN| + |CMN| = |ACMN|,$$

which implies the required inequality (50).

Finally, the points D' and E' may not coincide with M and N unless the original points D and E coincided with M and N. You can prove it (do) by backtracking moves 1 and 2.   □

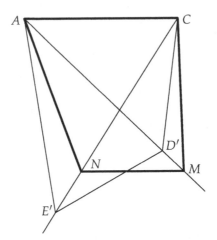

Figure 9.10

You can use this approach (do!) to find a nice, non-computational solution to problem ES2, as well.

10

Convex Figures and Erdős' Function $S_\alpha(F)$

Every book eventually reaches its ending (even *War and Peace* by Leo Tolstoy!). I thought Chapter 9 would conclude this book when I sent copies of the manuscript to four colleagues who wrote introductions for this book. Their input was most valuable; this is not just a polite thank you for their time. Their comments have aided me considerably in improving the book. You write a book in solitude, jumping from anxiety ("Nobody will like my baby!") to arrogance (as Michel de Montaigne put it about readership, "Few are enough for me; one will suffice, yea, less than one will content me").

It is comforting to come out of your cave, and to see that others understand you and enjoy your soon-to-be-born baby!

On December 29, 1989, Paul Erdős sent me a long letter filled with open problems. He opened it as follows:

> I looked fairly carefully through the whole book, and found it very interesting and in fact, it contained some material I did not know. Of course, my interests in geometry are of a more combinatorial nature. I did not know the nice result on $S(F)$.

Paul especially liked my function $S(F)$. "You very nicely solved $S(F)$ for $1/4$," he wrote, and immediately went on to generalize $S(F)$.

A. Soifer, *How Does One Cut a Triangle?*, Second Edition, DOI 10.1007/978-0-387-74652-4_10, 121
© Alexander Soifer 2009

Definition 10.1. (P. Erdős) Given a number α, $0 < \alpha < 1$, and a convex figure F of area 1. Then $S_\alpha(F)$ denotes the minimal positive integer n such that among any n points located inside or on the boundary of F there are three points that form a triangle of area α or less.

The main result of Chapter 8 can now be reformulated as follows:

Problem 10.2. For any convex figure F,

$$S_{1/4}(F) = 5 \text{ or } 6.$$

Paul Erdős then posed the following open problems, and offered ten-dollar prizes for their first solutions. With Paul's passing, I will pay these prizes on his behalf.

Problem 10.3. (P. Erdős) Find the smallest and the largest values of α such that for any convex figure F,

$$S_\alpha(F) = 5 \text{ or } 6.$$

Problem 10.4. (P. Erdős) Given α; let the smallest value of $S_\alpha(F)$ be attained for $F = F_1$ and the largest value of $S_\alpha(F)$ be attained for $F = F_2$.
  Can the difference

$$S_\alpha(F_2) - S_\alpha(F_1)$$

be arbitrarily large as we vary α?
  Do all intermediate values between $S_\alpha(F_1)$ and $S_\alpha(F_2)$ occur for some convex figures F?

Problem 10.5. (P. Erdős) Is there an α such that for all convex figures F of area 1 the function $S_\alpha(F)$ takes on just one value?

I would certainly like to see the following problem solved:

Problem 10.6. For every α and positive integer n, solve the following equation:

$$S_\alpha(F) = n,$$

i.e., classify (describe) all convex figures F for which this equality takes place.

Easier asked than solved! This problem appears to be hopeless in full generality. What about solving it for some "nice" values of α? The One-Hundred-Dollar Problem (Problem 8.6.6) asks you to solve Problem 10.6 for $\alpha = \dfrac{1}{4}$. Another nice value of α for which I would like to see this problem solved is $\alpha = \dfrac{3\sqrt{3}}{4\pi}$. Why?

Because Blaschke's result (Problem 8.7.5) can be reformulated as follows:

Problem 10.7. (W. Blaschke) Let F be a convex figure.

If $\alpha < \dfrac{3\sqrt{3}}{4\pi}$, then $S_\alpha(F) \geq 4$.

If $\alpha = \dfrac{3\sqrt{3}}{4\pi}$, then $S_\alpha(F) = 3$ only if F is an ellipse.

Finally, Paul Erdős suggests looking into polygons (rather than triangles) formed by points inside convex figures. Let us try.

Definition 10.8. Given a number α, $0 < \alpha < 1$; a positive integer n, $n \geq 3$; and a convex figure F of area 1. $S_\alpha^n(F)$ will denote the minimal positive integer m such that among any m points located inside or on the boundary of F, there are n points whose convex hull has area of at most α.

We can now reformulate Sas's result (Problem 8.7.6) as follows:

Problem 10.9. (Sas) Let F be a convex figure and n a positive integer, $n \geq 3$.

If $\alpha < \dfrac{n}{2\pi} \sin \dfrac{2\pi}{n}$, then $S_\alpha^n(F) > n$.

If $\alpha = \dfrac{n}{2\pi} \sin \dfrac{2\pi}{n}$, then $S_\alpha^n(F) = n$ only if F is an ellipse.

Many exciting problems can be posed about the function $\mathbf{S}_\alpha^n(F)$. Problems 10.4, 10.5, and 10.6 are good examples of such questions.

Now you, my reader, are ready to create and solve your own new and exciting problems.

Do!

And let me hear from you!

Part II
Developments of the Subsequent 20 Years

Since its release in early 1990, *How Does One Cut a Triangle?* has lived its own life, quite separate from mine. It was well received by all of the journals that reviewed it. More importantly, it inspired new solutions and new research by some wonderful mathematicians, young and not-so-young. You can see its influence in such papers as [L2], [L3], [L4], [HGL], [Ka1], and [Ka2].

Let us look at some of these results, which continue trains of mathematical thought you have read in Part I.

An Alternative Proof
of Grand Problem II

Let me add to the story you read at the end of Chapter 2 an alternative solution to Grand Problem II.

Ivan V. Arzhantsev, a.k.a. Vania, is currently an associate professor in the Department of Higher Algebra at Moscow State University. I am looking at Vania's original paper as I write these lines. It is (too) long and witty. Its idea is, however, simple. Let me quote a concise version that belongs to the pen of the Russian problem-solving legend and de facto chair of the Soviet Union National Mathematical Olympiad (the formal chair was A. N. Kolmogorov), Nikolaj (Kolya) N. Vasiliev, who left us so untimely. Kolya writes in a June 28, 1990, letter to me (I am translating his letter from the Russian for you and adding references to the required parts of the first solution):

Dear Sasha!

Your problem we have published... I think that the reasoning of this second [Vania's] solution is reduced simply to the following:

Let A, B, C be the angles of our triangle T, where A is much (i.e., more than 10 times) greater than B, and B is much greater than C. Assume that T is cut into 5 triangles similar to each other, and consequently some of its angles are partitioned into smaller angles. Denote by A_1, B_1, C_1 the largest parts of the partitions of the angles A, B, C, respectively. Since

A. Soifer, *How Does One Cut a Triangle?*, Second Edition, DOI 10.1007/978-0-387-74652-4_11,
© Alexander Soifer 2009

every angle of T is partitioned at most into 5 parts, it is obvious that $A_1 > B_1 > C_1$. Therefore, A_1, B_1, C_1 are the angles of the 5 triangles of the partition, their sum is $180°$ and thus they are equal to A, B, C, respectively, and the angles of T were actually not partitioned but rather cut off.

Now Figures 2.7 and 2.8 are the only options, and the reasoning accompanying these figures above completes the proof.   □

Kolya then informs me that my algebraic solution (presented above) and Vania's solution in Kolya's concise presentation will appear in *Kvant* magazine. They did, in December 1990 [SV].

Miklós Laczkovich on Cutting Triangles

This story—as many of my stories—begins with Paul Erdős.

In March 1989, Paul Erdős visited me in Colorado Springs for the first time. When our weeklong work was done, Paul asked me to drive him to Boulder to the house of Jan Mycielski, our mutual friend—and a fine mathematician with the broadest mathematical repertoire. The moment we sat down in Mycielski's house (or were we still standing?), Jan shocked us with the news: the young Hungarian mathematician Miklós Laczkovich had just solved the famous Tarski's Circle-Squaring Problem. Paul Erdős replied with delight and surprise, "I knew he [Laczkovich] was clever, but I did not know that he was *that* clever."

In 1925, the great Polish-turned-American logician and mathematician Alfred Tarski posed his celebrated problem:

Tarski's Circle-Squaring Problem 12.1. (A. Tarski, 1925 [T]) Is a (circular) disk equidecomposable to a square, i.e., can a disk be decomposed into finitely many pieces that can be rearranged to form a square?

Previously, another great Polish mathematician, Stefan Banach, and Tarski proved that in the 3-dimensional space $\mathbf{R}^3$ any two bounded sets with a non-empty interior are equidecomposable. This, of course, inspired Tarski's Circle-Squaring Problem, which

A. Soifer, *How Does One Cut a Triangle?*, Second Edition, DOI 10.1007/978-0-387-74652-4_12, 129
© Alexander Soifer 2009

stood open for 64 years until Miklós settled it in the positive. More-over, he proved a stronger result:

Laczkovich's Theorem 12.2. [L1] A (circular) disk is equidecompos-able into a square. Moreover, there is a decomposition of the disk into finitely many pieces that can be moved by translations alone to form the square.

I was most impressed by Mikós's accomplishment, too. Follow-ing the dinner with Jan Mycielski, his wife and Paul Erdős, I sent my congratulations to Laczkovich in Hungary, and received in return—and am looking at it, as I write these lines—an inscribed preprint of Miklós's celebrated proof. It was only natural that in early 1990 I sent to Miklós a return present: a copy of this book, which had just come off the printer's press. Little did I expect that it would inspire Miklós to do research in the direction of this book! I learned about it his June 24, 1992, letter:

> Dear Professor Soifer,
>   First of all, I would like to thank you for your beautiful book *How Does One Cut a Triangle?* This book motivated me to complete a paper about tilings of triangles, which I am sending in a separate cover. I submitted it to *Discrete Mathe-matics*.
>   With best regards,
>   Miklós Laczkovich

Unlike *Geombinatorics*, a quarterly founded by Branko Grün-baum and me in 1991, most journals are quite slow to publish their submissions. It took over 2 years for Miklós to see his proofs, as he informed me in a November 3, 1994, letter:

> Dear Alexander,
>   My paper "Tilings of triangles" will appear in *Discrete Mathematics* 137 (1995). I have seen the proofs, but I didn't receive reprints yet. I will send one as soon as I get them.
>   As I am not quite sure whether I gave you these reprints, I enclose "Tilings of the square with similar rectangles" (to

appear in the Fejes Tóth volume), and "Decomposition of convex figures into similar pieces" (to appear in *Discrete and Comp. Geometry*).

    With best wishes,
    Miklós

The reprint of "Tilings of triangles" arrived in 1995, 3 full years after the original preprint.

By a movie-like coincidence, I soon received another copy of this article—from *Zentralblatt für Mathematik*—and wrote a most complimentary review:

> In part 1 of his book *How Does One Cut a Triangle?* (Center for Excellence in Mathematical Education, Colorado Springs, 1990) the reviewer was interested in finding all integers n such that every triangle T can be dissected into n triangles similar to each other (or congruent to each other). The author here continues in this direction with a special interest in the kinds and quantity of similar (or congruent) to each other triangles Δ into which the triangle T can be dissected.
>
> Given a non-equilateral triangle T. If T can be dissected into triangles that are all similar to one triangle Δ, which is not similar to T, then we include Δ in the set $S(T)$. The author proves that $S(T)$ contains at most 6 non-similar triangles.
>
> Given a triangle T. If T can be dissected into triangles that are all congruent to one triangle Δ, which is not similar to T, then we include Δ in the set $C(T)$. The author proves that there are infinitely many isosceles triangles T such that $C(T)$ includes at least 6 non-similar triangles. Finally, the author studies dissections of equilateral triangles.
>
> In the process, the author discovers a good number of ingenious dissections.

Presently, I told Miklós that I was preparing an expanded edition of this book for Springer and offered him the opportunity to contribute. In his March 23, 2008, reply, Miklós generously agreed to contribute to this edition:

I also have several unpublished results (who doesn't)? I will try to send you a short (unpublished) note about the possible number of pieces of decomposing a triangle into congruent triangles; I believe this result would fit in your book nicely.

Indeed, on March 27, 2008, I received Miklós's note "On the number of pieces in tilings of triangles" [L4]. It is written in the highest quality academic style. I would like to elaborate, make the narration self-contained, and add illustrations so that my reader can more easily "digest" this fine work by Miklós Laczkovich. The ideas are all Miklós's from here to the end of this chapter!

Paul Erdős's Problem 6.7 provides the inspiration and the starting point. Erdős asks us to find all integers n for which there are triangles T and Δ such that T can be dissected into n triangles congruent to Δ. If we impose an additional condition requiring T and Δ be similar to each other, then we obtain the so-called *replicating tilings*, or *reptiles*. In 1964, Solomon W. Golomb [Go] proved the following existence result.

Golomb's Reptiles 12.3. [Go] There are reptiles with $n = k^2, 3k^2$, or $k^2 + m^2$ pieces.

Proof. Figure 2.2 delivers a proof for $n = k^2$.

We are positively lucky to have a right triangle with 30°–60°–90° angles, for it can be cut into 3 congruent triangles similar to it (Figure 12.1). Each of them can then be cut into k^2 congruent triangles as in Figure 2.2, and we get $n = 3k^2$ congruent triangles.

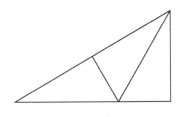

Figure 12.1

Finally, to get $n = k^2 + m^2$ congruent triangles, we construct a right triangle with legs k and m, divide it into two triangles by the altitude drawn to its hypotenuse, and finish by cutting the two triangles into k^2 and m^2 congruent triangles, respectively (Figure 12.2).

□

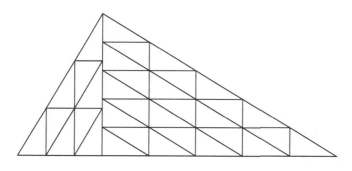

Figure 12.2

Decades later, in 1991, S. L. Snover, C. Waiveris and J. K. Williams [SWW] showed that no other value of n occurs as the number of pieces in a triangle-reptile.

If we remove the condition of similarity of T and Δ, then new values of n appear. For example, it is easy to see that an equilateral triangle can be dissected into $6k^2$ congruent triangles. All these numbers are of the form qm^2, where q is square-free, and if p is a prime divisor of q, then $p = 2$, $p = 3$ or $p = 4k + 1$.

In his fine 1995 paper (inspired by this book), Miklós Laczkovich showed ([L2], Theorem 3.4) that in the dissections of the equilateral triangle, infinitely many other values of n occur; namely, whenever P is a finite set of primes, the equilateral triangle can be dissected into n congruent triangles, where every $p \in P$ has an odd exponent in the prime factorization of n.

In order to present Laczkovich's new unpublished theorem, which appears in print for the first time here, we need to rely on his result from the same paper [L2]:

Tool 12.4. ([L2], Theorem 2.3) Let T be an isosceles triangle with angles $\alpha, \alpha, \pi - 2\alpha$ such that $\alpha \leq \pi/3$ and $\cos\alpha$ is rational. Then T can be dissected into congruent triangles with angles $\alpha, 2\alpha, \pi - 3\alpha$.

Now we are ready for the main result:

Laczkovich's Theorem 12.5. [L4] For every positive integer k, there are an integer m and triangle T such that T can be dissected into km^2 congruent triangles.

Proof. Let $1 < r < 2$ be a rational number, and let $\alpha = \cos^{-1}(r/2)$. Then $0 < \alpha < \pi/3$, and $\cos\alpha$ is rational. Let T denote the isosceles triangle with angles $\alpha, \alpha, \pi - 2\alpha$. Let the sides of T be $a \cdot \sin\alpha, a \cdot \sin\alpha$ and $a \cdot \sin(2\alpha)$. Since $\sin(2\alpha)/\sin\alpha = 2\cos\alpha$ is rational, we may choose a such that the sides of T are rational (we may take $a = 1/\sin\alpha$).

By Tool 12.4, T can be dissected into congruent triangles with angles $\alpha, 2\alpha, \pi - 3\alpha$. Let the sides of the pieces be $b \cdot \sin\alpha, b \cdot \sin(2\alpha)$ and $b \cdot \sin(3\alpha)$. Since $\sin(3\alpha)/\sin\alpha = 4\cos^2\alpha - 1$ is rational, the sides of the pieces are commensurable. Now the sides of T are packed with sides of the pieces, and thus the sides of the pieces are also rational. It follows that a/b is rational.

The number of pieces n of the dissection equals the ratio of the areas; that is,

$$
n = \frac{a^2 \cdot \sin^2\alpha \cdot \sin(2\alpha)}{b^2 \cdot \sin\alpha \cdot \sin(2\alpha) \cdot \sin(3\alpha)} = \frac{a^2/b^2}{4\cos^2\alpha - 1} = \frac{a^2/b^2}{r^2 - 1}
$$
$$
= (r^2 - 1)\left(\frac{a}{b(r^2 - 1)}\right)^2.
$$

We have proven the following: for every rational number $1 < r < 2$, there is a triangle T that can be dissected into $n = (r^2 - 1)c^2$ congruent triangles, where c is rational.

Now we turn to the proof of the theorem. Let k be given; we may assume that $k > 3$ and k is square-free. Let $r = (k+1)/(k-1)$. Then $r^2 - 1 = 4k/(k-1)^2 = kd^2$, where d is rational. Therefore, we can find a triangle T that can be dissected into $n = kc^2$ congruent triangles, where c is rational. Since k is square-free, it follows that n/k must be a square, which completes the proof.   □

One statement at the end of the introduction to Laczkovich's [L2] attracted my attention. Today, on May 5, 2008, Miklós clarified the problem for me and shared his thoughts about it in an e-mail.

Open Problem 12.6. (Laczkovich, 1995) Is there a positive integer n such that for every integer $N > n$ an equilateral triangle can be dissected into N congruent triangles?

Miklós informs me that he believes that the answer to this question is "no."

13

Matthew Kahle on the Five-Point Problem

I first met Matt on April 17, 1987. He was an 8th-grader, and his mom brought him to the Fourth Colorado Mathematical Olympiad. Matt was holding in his hands his baby sister. A month later I met the three in Washington, D.C., where Matt was on the Colorado team for National Competition of MATHCOUNTS (and I was a national judge). The following year, in 1988, Matt won third prize in the Colorado Mathematical Olympiad. His grades in school were low; even his geometry grade was a C. I asked his fine math teacher Judy Williamson,

—Why did Matt get a C in your geometry class?

—Because he does homework only when he likes problems. I have 40 students, and cannot possibly provide individual instruction.

I had no answer for Judy, but did have one for Matt. I told him:

—Any time you need a math problem, or simply to talk, come to me.

And Matt came: he took my university problem solving class while still a high school freshman. He was destined to do better in the Olympiad than his 1988 third prize: Matthew won first prize

A. Soifer, *How Does One Cut a Triangle?*, Second Edition, DOI 10.1007/978-0-387-74652-4_13, 137
© Alexander Soifer 2009

and a gold medal in both the 1990 and 1991 Colorado Mathematical Olympiads, and proved that he was the best high school mathematician in Colorado at the time. However, this did not open the doors of my University of Colorado for him. The reason was a low grade point average (about C-).

What was I to do? I wrote a letter to the Director of Admissions and Records of my university. I argued:

> We, the University of Colorado, need Matt more than he needs us. One day we would be proud to have such a brilliant alumni.

All this was to no avail. Randy Koouba, Director of Admissions and Records, replied:

> We have rules. Let him go to Pikes Peak Community College and improve his grade point average. Then he could try us again.

"Then" has not happened, as Colorado State University at Fort Collins offered Matt a full scholarship and Matt got his Bachelor's and Master's degrees there. Matt then earned his Ph.D. in mathematics from the University of Washington, Seattle. He is now a postdoctoral fellow at Stanford University.

My university—more precisely Chancellor Pam Shockley and Arts and Sciences Dean Tom Christensen—eventually approved an "admission window," i.e., exceptions in admissions for students who won any medal in Colorado Mathematical Olympiad during their high school career. In this, the university matched exceptions it long granted to American Olympic sportsmen. My thesis that we ought to reward brains at least as much as we reward arms and legs has finally celebrated its victory!

More than any contestant, Matt absorbed the Olympiad spirit. He competed in five Olympiads, and he came back seven times to serve as the Olympiad's judge, to participate in passing the flaming Olympic torch to the next generations of young mathematicians. In order to help with judging, Matt traveled 120 miles from Fort

Collins—you see him here in a 2001 photograph during one of such visits. In 2008, Matt traveled 1200 miles each way to come from Stanford to Colorado Springs.

Matthew Kahle (right) and Alexander Soifer on April 21, 2001,
during the XVIII Colorado Mathematical Olympiad.

In his essay [Kah] that appears in *Geombinatorics* in January 2009, Matt speaks of what he learned from me:

> Perhaps the most important thing that I learned from him is that we are free to ask our own mathematical questions and pursue our own interests, that we can trust our sense of aesthetics and our intuition. This was a very empowering idea to me, particularly when I was young and struggling with boredom and frustration with school.

Matt then talks about the Five-Point Problem (7.2.1) from this book:

> One problem which I have always remembered for its simple statement and somewhat tricky proofs ... appeared on the Colorado Math Olympiad in 1988. It stayed with me, and I have occasionally sat down to see a new solution to it. . .
>
> One wonders if this is the best possible result, and in a sense it is: it is easy to check that "five" cannot be replaced by "four." But what about the 1/4?
>
> I must admit that I did not make any real progress toward proving [Problem 7.2.1] during the Olympiad, but now I will have a (very) small revenge on the problem by improving $1/4 = 0.25$ to $6/25 = 0.24$. In the spirit of Sasha Soifer and his wonderful journal *Geombinatorics*, we leave the reader with some open problems as well.

And so Matt proves the following result, improving on the result of Problem 7.2.1:

Kahle's Theorem 13.1. [Kah] Any 5 points in a triangle of area 1 contain 3 points that form a triangle of area at most $6/25$.

Matt uses a search for areas inside the given triangle where the points may be, very much like my proof of Problem 7.2.1. Instead of a triangular grid with 16 triangles (Figure 7.6), Matt uses a 100-triangle grid. His proof is much more complex, and I recommend you read it in *Geombinatorics*.

Having taken stock in the job well done, Matt formulates a natural question:

Open Problem 13.2. [Kah] Find the minimum α such that any 5 points in a triangle of area 1 contain 3 points that form a triangle of area at most α.

Matt observes that the minimum exists due to compactness discussed in Boltyanski–Soifer book [BS], and finds the first lower bound:

Lower Bound 13.3. [Kah] $\alpha \geq 1/6$.

Proof. Behold (Figure 13.1):

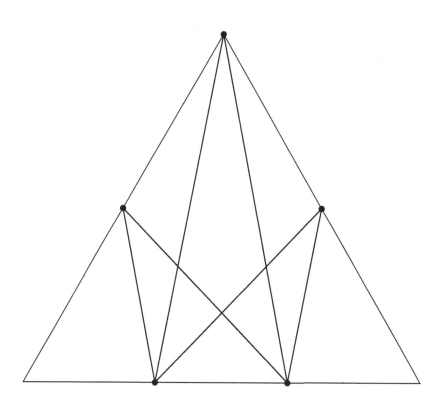

Figure 13.1 Points *b* and *e* are midpoints of the sides, while *c* and *d* partition the base into three equal parts. Triangles *abc*, *ade*, *bcd*, and *cde* all have area exactly 1/6.

$\square$

Of course, Kahle's Theorem provides the upper bound:

$$\alpha \leq 6/25.$$

Now it is time for conjecturing the exact value of α—and Matt obliges:

Kahle's Conjecture 13.4. [Kah] $\alpha = 1/6$.

Matthew Kahle's article [Kah] is dedicated to my round-numbered birthday. Thank you, Matt, your work and you have made my day!

Now back to work everyone, for Matt assigned homework to all of us, he himself included.

Soifer's One-Hundred-Dollar Problem and Mitya Karabash

In 2005, the brilliant young freshman from Columbia University Mitya Karabash came to the University of Colorado at Colorado Springs, where I supervised his summer research. We looked at problems of tiling and covering, the chromatic number of the plane, and problems posed by Paul Erdős and me in this book, and wrote a joint research paper. Since then, we corresponded and wrote more joint papers. Mitya visited me several times at Princeton University. Mitya worked for years on the One-Hundred-Dollar Problem from this book, and we discussed his progress. Finally, on June 4, 2007, on the blackboard of my 7th floor Princeton office, Mitya showed me the most important result related to this book in the 18 years that have passed since its birth in 1990.

What exactly did Mitya prove? He did not solve the One-Hundred-Dollar Problem (Problem 8.6.6) that asks to classify all convex figures F for which $\mathbf{S}(F) = 6$. But he settled Conjecture 8.6.7 in the negative by constructing a figure F such that $\mathbf{S}(F) = 6$, yet F cannot be mapped by an affine transformation into a however-narrow frame formed by two concentric regular pentagons.

Mitya's construction is too dense to be included here. It recently appeared as a sequence of two papers, [Ka1] and [Ka2], in the quarterly *Geombinatorics*.

Mitya creates an elaborate *Construction Tool* and then proves the following theorem:

A. Soifer, *How Does One Cut a Triangle?*, Second Edition, DOI 10.1007/978-0-387-74652-4_14, 143
© Alexander Soifer 2009

Karabash's Theorem 14.1. ([Ka1]) The class of figures F satisfying $S(F) = 6$ (which Mitya calls "the Soifer class") is equivalent to the class of figures that can be obtained by the Construction Tool.

The structure of the figures F with $S(F) = 6$ appears to be even more complex than I imagined twenty years ago—in fact, so complex that Mitya has formulated the following conjecture:

Karabash's $100 Conjecture 14.2. ([Ka2]) It is impossible to classify figures in the set $S^{-1}(6)$.

Of course, it depends what "classify" means. In a definite sense, Theorem 14.1 is a classification, but not one we can easily use to determine whether a given figure F belongs to the class of interest. Observant readers would now ask, quite appropriately, what does "given" mean?

As a result of Mitya's impressive papers, we do not even have a working conjecture for the One-Hundred-Dollar Problem. So, in order to solve the One-Hundred-Dollar Problem, you need to come up with a conjecture—and then prove it.

Even though Mitya has not solved the One-Hundred-Dollar Problem, I wrote a $50 check to him for a major advance. Mitya asked instead for an inscribed copy of this book, which he received in my Princeton home on June 4, 2007. You can see this moment depicted in the photograph that follows.

Mitya was invited to start in the fall of 2008 in Ph. D. programs at Harvard, MIT, and other premier universities. He has chosen to write his thesis on probability theory at the Courant Institute of Mathematical Sciences at the New York University. I hope he will continue to play with geometry and topology as well.

Mitya Karabash on June 4, 2007, in Princeton, receiving this book for advancing the One-Hundred-Dollar Problem. Photograph by Maya Soifer.

Coffee Hour and the Conway–Soifer Cover-Up

During the years 2002–2004 I was visiting Princeton University with its fabulous mathematics department, a great fixture of which was a daily coffee hour in the commons room, attended by everyone, from students to the Beautiful Mind (John F. Nash, Jr.). For one such coffee hour I came thinking again about the drawing depicted in this book in Figure 2.2. This time I imagined that we dealt with equilateral triangles, and the crux of the figure was a demonstration that n^2 unit triangles can cover a triangle of side n.

I asked myself a question where the continuous clashed with the discrete: if we were to enlarge the side length of the large triangle from n to $n + \varepsilon$, how many unit triangles would we need to cover it? This comprised a new open problem:

Cover-Up Problem 15.1. Find the minimum number of unit equilateral triangles required to cover an equilateral triangle of side $n + \varepsilon$.

During the next coffee hour, I posed the problem to a few Princeton colleagues. It immediately excited John H. Conway, the John von Neumann Professor of Mathematics. From the commons room he went to the airport to fly to a conference. On board the airplane, John found a way to do the job with just $n^2 + 2$ unit triangles! (Area considerations alone show the need for at least $n^2 + 1$ of them.) Conway shared his cover-up with me upon his return—at a coffee hour,

A. Soifer, *How Does One Cut a Triangle?*, Second Edition, DOI 10.1007/978-0-387-74652-4_15, 147
© Alexander Soifer 2009

John H. Conway (left) and Alexander Soifer,
July, 2007. Photograph by Simon Kochen.

of course. Now it was my turn to travel to a conference, and have quality time on an airplane. What I found was a totally different cover-up with the same number, $n^2 + 2$ unit triangles!

Upon my return, at a coffee hour, I shared my cover-up with John Conway. We decided to publish our results together. John suggested setting a new world record for the least number of words in a paper, and submitting it to the *American Mathematical Monthly*. On April 28, 2004, at 11:50 AM (computers record the exact time!), I submitted our paper that included just two words, "$n^2 + 2$ can," and our two drawings. I am compelled to reproduce our submission here in its entirety.

**Can $n^2 + 1$ unit equilateral triangles cover
an equilateral triangle of side $> n$, say $n + \varepsilon$?**

John H. Conway & Alexander Soifer
Princeton University, Mathematics
Fine Hall, Princeton, NJ 08544, USA
conway@math.princeton.edu   asoifer@princeton.edu

$n^2 + 2$ can:

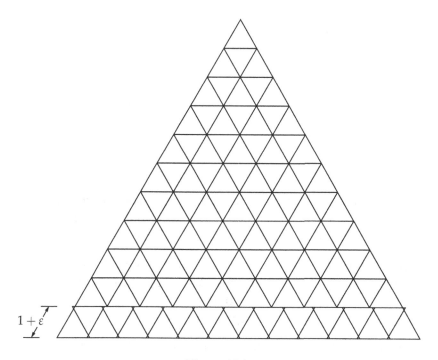

Figure 15.1

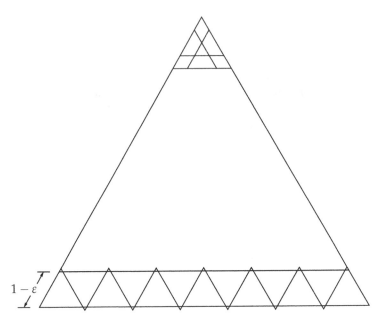

Figure 15.2

The American Mathematical Monthly was surprised, and did not know what to do about our 2-word article. Two days later, on April 30, 2004, the Editorial Assistant, Margaret Combs, wrote:

> The Monthly publishes exposition of mathematics at many levels, and it contains articles both long and short. Your article, however, is a bit too short to be a good *Monthly* article. . . A line or two of explanation would really help.

The same day at the coffee hour I asked John, "What do you think?" His answer was concise, "Do not give up too easily." Accordingly, I replied to *The Monthly* the same day:

> I respectfully disagree that a short paper in general—and this paper in particular—merely due to its size must be "a bit too

short to be a good *Monthly* article." Is there a connection between quantity and quality?... We have posed a fine (in our opinion) open problem and reported two distinct "behold-style" proofs of our advance on this problem. What else is there to explain?

The Monthly apparently felt outgunned, for on May 4, 2004, the reply came from *The Monthly*'s top gun, Editor-in-Chief Bruce Palka:

> *The Monthly* publishes two types of papers: "articles," which are substantive expository papers ranging in length from about six to twenty-five pages, and "notes," which are shorter, frequently somewhat more technical pieces (typically in the one-to-five page range). I can send your paper to the notes editor if you wish, but I expect that he'll not be interested in it either because of its length and lack of any substantial accompanying text... The standard way in which we use such short papers these days is as "boxed filler" on pages that would otherwise contain a lot of the blank space that publishers abhor... If you'd allow us to use your paper in that way, I'd be happy to publish it.

John Conway and I accepted the "filler," and our paper [CS1] was published in the January 2005 issue. *The Monthly*, however, invented a title without any consultation with the authors, and added our title to the body of the article!

We also ran our little article in *Geombinatorics*, where we additionally observed that the "equilaterality" is essential, for otherwise $n^2 + 1$ triangles, similar to the large triangle T and with the ratio of sizes $1 : n + \varepsilon$, can cover T (Figure 15.3).

Mitya Karabash then joined me for further explorations of this problem. First of all, we observed the following result, which is better than its simple proof:

Non-Equilateral Cover-Up 15.2. [KS1] For *every* non-equilateral triangle T, $n^2 + 1$ triangles similar to T and with the ratio of sizes $1 : (n + \varepsilon)$, can cover T.

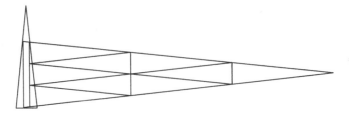

Figure 15.3

Proof. An appropriate affine transformation maps the equilateral triangle and its covering shown in Figure 15.2 into T. This transformation produces a covering of T with $n^2 + 2$ triangles similar to T. We can now cover the top triangle with 2 tiling triangles instead of 3 as shown in Figure 15.4, thus reducing the total number of covering triangles to $n^2 + 1$.   □

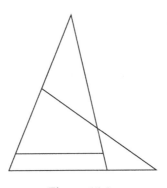

Figure 15.4

Mitya and I then generalized the problem from covering a triangle to covering much more complex figures we named *trigons*.

An *n-trigon* T_n is the union of n triangles from the standard triangulation of the plane such that a triangular rook can find a path between any two triangles of T_n, i.e., the union of n edge-connected triangles. If the triangulation is equilateral, then we say that the *n*-

trigon is *equilateral*. You can see an example of an equilateral 9-trigon in Figure 15.5.

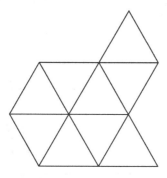

Figure 15.5

In our cover-up games, we assume that the ratio of the corresponding sides of the triangles forming the trigon and the tiling triangles is $(1 + \varepsilon)/1$. We proved an important result:

Karabash–Soifer's Trigon Theorem 15.3. [KS1] An n-trigon T_n can be covered with

1) $n + 2$ triangles if the trigon is equilateral;
2) $n + 1$ triangles if the trigon is non-equilateral.

In spite of all the progress, however, one "little" question remains open. I will formulate it here as a conjecture (because John Conway and I thought we knew the answer—we just had no idea how to prove it):

Cover-Up Conjecture 15.4. (Conway–Soifer, 2004) An equilateral triangle of side $n + \varepsilon$ cannot be covered by $n^2 + 1$ unit equilateral triangles.

Right after Cover-Up Problem 15.1, I created the **Cover-Up Squared Problem**. Naturally, a square of side n can be covered by n^2 unit squares. When, however, I let the side length increase merely to $n + \varepsilon$, I got a new open problem:

Cover-Up Squared Problem 15.5. [S4] Find the smallest number $P(n)$ of unit squares that can cover a square of side length $n + \varepsilon$.

I devised a covering approach illustrated in Figure 15.6.

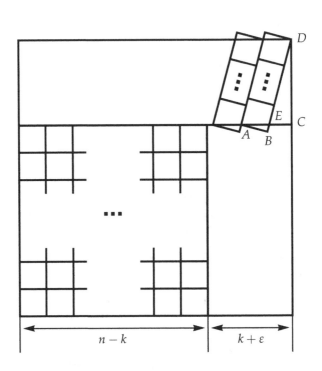

Figure 15.6

My results were followed by the joint ones by Mitya Karabash and me. The best Mitya and I were able to do in the cover-up squared was to match Paul Erdős and Ronald L. Graham's dual 1975 result [EG1] on packing squares in a square:

Karabash–Soifer's Theorem 15.6. [KS2] $P(n) = n^2 + O(n^{7/11})$.

Let me explain the "big O" notation. We write $f(n) = O(g(n))$ if asymptotically the function $f(n)$ grows not faster than a constant multiple of $g(n)$.

Immediately upon the publication of this result, on July 18, 2008, I received an e-mail from Ronald L. Graham, saying, "I received the latest issue of Geombinatorics with your nice article on square covering." I mention this e-mail because, like in a movie made for television, there was a continuation to Ron's e-mail. A mere month later, he sent me a manuscipt [CG], jointly written with Fan Chung, entitled "Packing equal squares into a large square." In it, they improved the 33-year-old Graham–Erdős result [EG1], and also Mitya's and my Theorem 15.6:

Chung–Graham's Theorem 15.7. [CG]

$$P(n) = n^2 + O\left(n^{(3+\sqrt{2})/7} \log n\right).$$

Do you see why the Chung–Graham result is slightly better than Karabash–Soifer? Here is why:

$$7/11 = 0.636363\ldots$$

$$\left(3 + \sqrt{2}\right)/7 = 0.630601\ldots$$

Would you like to hear "ttha rrrest of the story," as Paul Harvey used to say on the radio? On November 5, 2008, I was asked by the Editor-in-Chief Ole Warnaar to referee the Chung–Graham manuscript for the *Journal of Combinatorial Theory, Series A*. On December 12, 2008, I wrote my report:

REVIEW OF CHUNG–GRAHAM MS

Dear Professor Warnaar,

I have read the manuscript with great interest and pleasure. The results represent a small but important improvement, attained by clever improvements in the approach of Erdős–Graham 1975...

I enthusiastically recommend this article for publication in JCTA.

The Cover-Up Squared Problem remains open, both in search for the asymptotically lowest possible solution and for exact values for small n. Mitya and I conjecture:

Cover-Up Square Conjecture 15.8. [KS2] $P(n) = n^2 + \Omega(n^{1/2})$.

We write $f(n) = \Omega(g(n))$ if asymptotically the function $f(n)$ grows not slower than a constant multiple of $g(n)$.

16

Farewell to the Reader

Thank you for holding my book in your hands. I welcome your ideas, comments, conjectures, solutions of problems presented here and new problems you may create. They may inspire a new edition of this book. I hope we will meet again on the pages of my other books.

As Paul Erdős used to say at the end of his lectures, "everything comes to an end," and so has this book. However, if you are inclined to continue your explorations of mathematics with me, I have good news for you. This book is one of my seven books that Springer has or soon is going to publish.

If you are receptive to the visual appeal of geometry, you may wish to read the new expanded edition of *Geometric Etudes in Combinatorial Mathematics* [S9]. Its first edition [BS] was published in 1991. These books show how geometric insight does wonders in service to combinatorics. In addition, the books contains beautiful geometric version of some classical theorems of analysis.

If you are interested in a mixture of exciting problems, I recommend you to work through the new edition of the book *Mathematics as Problem Solving* [S7]. Its first edition [S1] came out in 1987.

You will find an even greater variety of problems in the new book *Colorado Mathematical Olympiad: The First 20 Years and Further Explorations* [S10]. *The First 10 Years* was published in 1994 [S3]. These books also offer 20 bridges from the problems of mathemat-

A. Soifer, *How Does One Cut a Triangle?*, Second Edition, DOI 10.1007/978-0-387-74652-4_16, 157
© Alexander Soifer 2009

ical olympiads to problems of "real" mathematics. You will find there open problems that could inspire you to start your own mathematical research!

Election Day, November 4, 2008 (the "yes-we-can" day) saw the release of the book I dreamed of and worked on for 18 year, *The Mathematical Coloring Book: Mathematics of Coloring and the Colorful Life of Its Creators* [S5]. This voluminous book offers a beautiful mathematics of coloring (so-called *Ramsey Theory*), historical investigations into lives of mathematicians, from the Nazi time in Germany to the devastated-by-World War II Netherlands. The history allowed me to pose questions which have not lost their urgency today, such as the role of a scholar in tyranny. The book presents the aesthetics of mathematics as an art, a philosophy of its foundations, and the psychology of mathematical and historical discovery. The Nobel Laureate Boris Pasternak [Pas] expressed my goals in this book better and more concisely than I could—great poets often do it well:

> *I bring here all: what have I lived thru,*
> *And that what keeps my soul alive,*
> *My rectitude and aspirations,*
> *And what have seen my own eyes.*

My next book will not include mathematics. However, a great twentieth century mathematician will be the hero of the book, which will therefore be entitled *Life and Fate: In Search of Van der Waerden* [S6]. I hope it will be published in 2010.

The book of open problems of the legendary mathematician Paul Erdős (1913–1996) will come next, likely in 2011: *Problems of PGOM Erdős* [ES]. I would not have attempted to write it, but in 1990 Paul asked me to join him in this endeavor, and thus it will be our joint book. As you may know, Paul Erdős was the greatest problem creator of all time. You will be able to work on his problems because no background knowledge is required to understand the majority of them. Moreover, many problems allow young mathematicians to advance and find partial solutions.

The book after Erdős will be either *The Art on the Frontier of Cultures: The Fang People of West Equatorial Africa and Their Neighbors*, or

Memory in Flashback. The former would be a result of my ongoing study of African Art and culture, inspired by the great anthropologist, my hero and friend James W. Fernandez. The latter will be a collection of humorous and noteworthy moments of my life, meetings with great people of many creative professions, and lessons from both sides of the Atlantic.

Having finished this book, you have become my alumnus, a title that carries responsibility to stay in touch, to send me your most enjoyable solutions, and your new problems, conjectures, suggestions and ideas. Rest assured I will always be delighted to hear back from you!

Appendix A

Cutting a Triangle into Congruent Triangles

Main Result of Chapter 4. If every triangle can be cut into k congruent triangles, then $\sqrt{k}$ is an integer.

Proof by S. Slobodnik. Let k be a number such that every triangle can be cut into k congruent triangles.

We define a sequence T_n of triangles as follows: For every n we pick a triangle T_n with the angles

$$A_n = \frac{\pi}{3}\left(1 - \frac{\sqrt{3}}{100n}\right)$$

$$B_n = \frac{\pi}{3}\left(1 - \frac{\sqrt{2}}{100n}\right)$$

$$C_n = \frac{\pi}{3}\left(1 + \frac{\sqrt{2}+\sqrt{3}}{100n}\right)$$

and its partition P_n into k copies of a triangle T_n'.

By Tool 2.1, the triangles T_n and T_n' are similar. Their corresponding sides are in the ratio $\sqrt{k} : 1$.

Let M_n and M_n' denote the longest sides of the triangles T_n and T_n', and m_n and m_n' denote the shortest sides of T_n and T_n'. Then, for the number i_n of sides of triangles T_n' lying on a certain side of the triangle T_n in the partition P_n (see Figure 3.1), we get the following inequality:

$$\frac{m_n}{M'_n} \le i_n \le \frac{M_n}{m'_n}.$$

By the Theorem of Sines, we can replace the ratios of sides by the ratios of sines of the corresponding angles:

$$\sqrt{k}\frac{\sin\left[\frac{\pi}{3}\left(1 - \frac{\sqrt{3}}{100n}\right)\right]}{\sin\left[\frac{\pi}{3}\left(1 + \frac{\sqrt{2}+\sqrt{3}}{100n}\right)\right]} \le i_n \le \sqrt{k}\frac{\sin\left[\frac{\pi}{3}\left(1 + \frac{\sqrt{2}+\sqrt{3}}{100n}\right)\right]}{\sin\left[\frac{\pi}{3}\left(1 - \frac{\sqrt{3}}{100n}\right)\right]}$$

As n increases without bound, the ratios of sines in the left and right sides of the inequality approach 1. Therefore,

$$\lim_{n \to \infty} i_n = \sqrt{k}.$$

But $\{i_n\}$ is a sequence of integers, so its limit $\sqrt{k}$ must be an integer as well!   □

Appendix B

The Five-Point Problem

Problem 7.2.1. Among any five points in a triangle of area 1, there are three points that form a triangle of area at most $1/4$.

Proof by C. Rousseau. Our proof uses Tool 7.1.3. The desired result follows from this tool once we have proven the following claim:

> *Among any five points in a triangle of area 1, there are three points that lie in a parallelogram of area at most $1/2$.*

Assume the contrary, i.e., that we can pick five points so that no three of them lie in such a parallelogram. The trick is now to choose the appropriate set of parallelograms that show that this assumption leads to a contradiction. There are many ways to do this; here is one. We begin by dividing the original triangle into four congruent triangles, A, B, C, D, by connecting the midpoints as shown below.

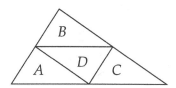

Let a, b, c, d denote the number of points lying in A, B, C, D, respectively (if a point lies on the common boundary of two triangles, we make an arbitrary choice of the triangle to which it belongs). Thus,

$$a + b + c + d = 5. \tag{1}$$

A *type I* parallelogram is the union of two of the four triangles and has area $1/2$, like the one shown below.

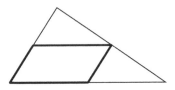

Type I Parallelogram

By considering the three type I parallelograms, we see that our assumption requires

$$a + d \leq 2$$
$$b + d \leq 2$$
$$c + d \leq 2.$$

Adding these inequalities and using (1), we find $2d \leq 1$. Since d is a whole number, we are forced to admit that $d = 0$. Now,

$$a + b + c = 5, \tag{2}$$

and we need to look at some more parallelograms. Divide each of the triangles A, B, C into four congruent triangles as shown below.

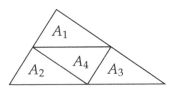

Triangle A

Let a_1, a_2, a_3, a_4 tell us how many of the five points lie in sub-regions A_1, A_2, A_3, A_4, respectively, of triangle A (boundary points are treated as before, so $a = a_1 + a_2 + a_3 + a_4$).

Define $b_1, \ldots, b_4$ and $c_1, \ldots, c_4$ in the same way. Now we introduce *type II* and *type III* parallelograms as shown below. Each of these parallelograms has area $1/2$. A type II parallelogram contains six of the twelve subregions $A_1, \ldots, C_4$ and a type III parallelogram contains four of these subregions.

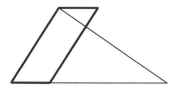

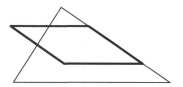

Type II Parallelogram   **Type III Parallelogram**

Since no parallelogram of area at most $1/2$ contains three or more of the five points, it is certainly clear that $a_1 + b_2 \leq 2$. In fact, $a_1 + b_2 \leq 1$, as we now show. By considering the appropriate type II and type III parallelograms, we find

$$a_1 + a_2 + a_4 + b_1 + b_2 + b_4 \leq 2$$
$$a_1 + b_2 + b_3 + b_4 \leq 2$$
$$a_1 + a_3 + a_4 + b_2 \leq 2.$$

If $a_1 + b_2 = 2$, then $a_2 = a_3 = a_4 = b_1 = b_3 = b_4 = 0$, so $a + b = 2$ and $c = 3$. Since having three points in triangle C has been disallowed, we can now assume $a_1 + b_2 \leq 1$. By symmetry, the same argument gives $b_3 + c_1 \leq 1$ and $c_2 + a_3 \leq 1$. Combining this information with the fact that no type II parallelogram contains three or more points, we have the following system of inequalities:

$$a_1 + a_2 + a_4 + b_1 + b_2 + b_4 \leq 2 \tag{3}$$
$$b_1 + b_3 + b_4 + c_1 + c_3 + c_4 \leq 2 \tag{4}$$
$$c_2 + c_3 + c_4 + a_2 + a_3 + a_4 \leq 2 \tag{5}$$
$$a_1 + b_2 \leq 1 \tag{6}$$
$$b_3 + c_1 \leq 1 \tag{7}$$
$$c_2 + a_3 \leq 1 \tag{8}$$

Adding inequalities (3)–(8) and using $a = a_1 + a_2 + a_3 + a_4$, etc., we obtain

$$2(a + b + c) \leq 9. \tag{9}$$

Now (2) and (9) give the desired contradiction and the proof is complete. $\qquad\square$

References

[B1]   Blaschke, W., Über affine geometrie III: Eine minimumeigenschaft der ellipse, *Berichte über die Verhandlungen*, B. G. Teubner, Leipzig, 1917.

[BG]   Boltjansky, V., and Gohberg, I., *Results and Problems in Combinatorial Geometry*, Cambridge University Press, Cambridge, 1985.

[BS]   Boltyanski, V. G., and Soifer, A., *Geometric Etudes in Combinatorial Mathematics*, Center for Excellence in Mathematical Education, Colorado Springs, 1994.

[CL]   Chakerian, G. D., and Lange, L. M., Geometric extremum problems, *Math Magazine* 44 (1971), 57–69.

[CG]   Chung, F., and Graham, R., Packing equal squares into a large square, *J. Combin. Theory, Ser. A* (to appear in 2010 or so).

[CS1]   Conway, J. H., and Soifer, A., Covering a Triangle with Triangles, *Amer. Math. Monthly* 112(1), 2005, 78.

[CS2]   Conway, J. H., and Soifer, A., Cover-Up, *Geombinatorics* XIV(1), 2004, 8–9.

[D1]   Dolciani, M. P., Graham, J. A., Swanson, R. A. and Sharrow, S., *Algebra 2 and Trigonometry*, Houghton Mifflin Company, Boston, 1989.

[Ei]   Einstein, A., On Freedom, in *Albert Einstein: Out of My Later Years*, Wings Books, New York, 1993.

[E1]   Erdős, P., Some of my old and new problems in elementary number theory and geometry, *Congressus Numerantium* 50 (1985), 97–106.

[E2]   Erdős, P., Some old and new problems in combinatorial geometry, *Annals of Discrete Mathematics* 20 (1984), 129–136.

[E3]   Erdős, P., *The Art of Counting: Selected Writings*, MIT Press, 1973.

[EG1]   Erdős, P., and Graham, R. L., On Packing Squares with Equal Squares, *J. Combin. Theory, Ser. A* 19 (1975), 119–123.

[EG2]   Erdős, P., and Graham, R. L., *Old and New Problems and Results in Combinatorial Number Theory*, L'Enseignement Mathématique, Université de Genève, 1980.

[ES]   Erdős, P., and Soifer, A., *Problems of PGOM Erdős*, Springer, New York, to appear.

[FS]   Fulton, C. M., and Stein, S. K., Parallelograms inscribed in convex curves, *American Mathematics Monthly* 67 (1960), 257–258.

[FT]   Fejes Tóth, L., *Lagerungen in der Ebene, auf der Kugel, und in Raum*, Springer, Berlin, 1953.

[Go]   Golomb, S. W., Replicating figures in the plane, *Math. Gaz.* 48 (1964), 403–412.

[G1]   Grünbaum, B., Measures of symmetry for convex sets, *Proc. Sympos. Pure Math.* 7 (Convexity), Providence (USA) (1963), 233–270.

[G2]   Grünbaum, B., Borsuk's problem and related questions, *Proc. Sympos. Pure Math.* 7 (Convexity), Providence (USA) (1963), 271–284.

[HDK]   Hadwiger, H., Debrunner, M., and Klee, V., *Combinatorial Geometry in the Plane*, Holt, Rinehart, and Winston, New York, 1964.

[HGL]   Hansheng, D., Gangsong, L., and Lin, S., On a sort of Heilbronn problem, *Discrete Math.*, submitted in September 2005.

[H]   Hofmannsthal, H. von, *Buch der Freunde*, Leipzig Insel, 1922. Translated by Douglas Robertson, http://shirtysleeves.blogspot.com/2008/04/translation-of-buch-der-freunde-by-hugo.html

[Kah]   Kahle, M., Points in a triangle forcing small triangles, *Geombinatorics* XVIII(3) (2009), 114–128.

[Ka1]   Karabash, D., On The Soifer Fifty Dollar Problem, Part I: Construction, *Geombinatorics* XVII(2) (2007), 68–77.

[Ka2]   Karabash, D., On The Soifer Fifty Dollar Problem, Part II: The Existence of the Counterexample to the Conjecture, *Geombinatorics* XVII(3) (2008), 124–128.

[KS1]   Karabash, D., and Soifer, A., On Covering of Trigons, *Geombinatorics* XV(1) (2005), 13–17.

[KS2]   Karabash, D., and Soifer, A., Note on Covering Square with Equal Squares, *Geombinatorics* XVIII(1) (2008), to appear.

[K1]   Klee, V., What is a convex set? *American Mathematical Monthly* 78(6) (1971), 616–631.

[K2]   Klee, V., Some unsolved problems in plane geometry, *Math. Magazine* 52(1) (1979), 131–145.

[L1]   Laczkovich, M., Equidecomposability and Discrepancy: A Solution of Tarski's Circle-Squaring Problem, *J. für die Reine und Angewandte Math.* 404 (1990), 77–117.

[L2]   Laczkovich, M., Tilings of triangles, *Discrete Math.* 140 (1–3) (1995), 79–94.

[L3]   Laczkovich, M., Tilings of Polygons with Similar Triangles, II, *Discrete Comput. Geom.* 19 (1998), 411–425.

[L4]   Laczkovich, M., On the Number of Pieces in Tilings of Triangles, A note written for this book, 2008.

[Lyu]   Lyusternik, L. A., *Vypuklye Figury i Mnogogranniki* (Convex Figures and Polyhedra) (Russian), Moscow, 1956. English Translation: *Convex Figures and Polyhedra*, Dover, New York, 1963.

[M1]   Montaigne, Michel de, *Essayes, John Florio's Translation*, The Modern Library, New York, 1933.

[MP]   Moser, W., and Pach, J., *100 Research Problems in Discrete Geometry*, Department of Mathematics and Statistics, McGill University, Montreal, Canada, 1986.

[Pas]   Pasternak, B., translated especially for *The Mathematical Coloring Book* [S5] by Ilya Hoffman.

[R1]   Radziszewski, C., Sur un probleme extremal relatif aux figures inscrites dans les figures convexes, *C. R. Acad. Sci Paris* 235 (1952), 771–773.

[Sa]   Sas, Az ellipszis egy extrem tulajdonsagarol (On an extremal property of the ellipse) (Hungarian), *Matematikai es fizikai lapok* 98 (1941), 523–542.

[SWW]   Snower, S. L., Waiveris, C., and Williams, J. K., Rep-tiling for triangles, *Discrete Comput. Geom.* 13 (1995), 569–572.

[S1]   Soifer, A., *Mathematics as Problem Solving*, Center for Excellence in Mathematical Education, Colorado Springs, CO, 1987.

[S2]   Soifer, A., *How Does One Cut a Triangle?* First ed., Center for Excellence in Mathematical Education, Colorado Springs, CO, 1987.

[S3]   Soifer, A., *Colorado Mathematical Olympiad: The First Ten Years and Further Explorations*, Center for Excellence in Mathematical Education, Colorado Springs, CO, 1987.

[S4]   Soifer, A., Cover-Up Squared, *Geombinatorics* XIV(4) (2005), 221–226.

[S5]   Soifer, A., *The Mathematical Coloring Book: Mathematics of Coloring and the Colorful Life of Its Creators*, Springer, New York, 2009.

[S6]   Soifer, A., *Life and Fate: In Search of Van der Waerden*, Springer, New York, to appear.

[S7]   Soifer, A., *Mathematics as Problem Solving*, 2nd edition, Springer, New York, 2009.

[S8]   Soifer, A., *How Does One Cut a Triangle?* 2nd edition, Springer, New York, 2009.

[S9]   Soifer, A., *Geometric Etudes in Combinatorial Mathematics*, 2nd edition, Springer, New York, 2009.

[S10]   Soifer, A., *Colorado Mathematical Olympiad: The First Twenty Years and Further Explorations*, Springer, New York, to appear.

[SV]   Soifer, A., and Vasiliev, N., Solution to Problem M1234, *Kvant* (1990), #12, 24–25.

[Sü]   Süss, W., Ueber Parallelogramme und Rechtecke, die sich ebenen Eibereichen einbeschreiben lassen, *Rend. Mat. e Appl. (5)* 14 (1955), 338–341.

[T]   Tarski, A., Problème 38, *Fundamenta Mathematicae* 7 (1925), 381–382.

[Y1]   Yaglom, I. M., *Kak Razrezat' Kvadrat?* (How Does One Cut a Square?) (Russian), Nauka, Moscow, 1968.

[Y2]   Yaglom, A. M., and Yaglom, I. M., *Neelementarnye Zadachi v Elementarnom Izlodgenii* (Non-Elementary Problems in Elementary Presentation) (Russian), GITTL, Moscow, 1954. English translation: *Challenging Mathematical Problems with Elementary Solutions*, Vol. I & II, Dover Publications, New York, 1987.

[Y3]   Shkljarski, D. O., Chenzov, N. N., and Yaglom, I. M., *Geometricheskije Ozenki i Zadachi iz Kombinatornoj Geometrii* (Geometric Estimates and Problems from Combinatorial Geometry) (Russian), Nauka, Moscow, 1974.

[Y4]   Yaglom, I. M., *Geometricheskije Preobrazovanija* (Geometric Transformations) (Russian), Nauka, Moscow, Vol. 1, 1955; Vol. 2, 1956. A partial English translation: *Geometric Transformations*, Mathematical Association of America, Washington, D.C., Vol. 1, 1962; Vol. 2, 1968; Vol. 3, 1973.

[YB]   Yaglom, I. M., and Boltjanski, V. G., *Vypuklye Figury* (Convex Figures), GITTL, Moscow, 1951. English translation: *Convex Figures*, Holt, Rinehart and Winston, New York, 1961.

Notation

	Page
$|a|$, $|AB|$	80
$|\vec{a}|$, $|\overrightarrow{AB}|$	80
$|F|$, $|ABC|$, $|ABCD|$	65
$\mathbf{S}(F)$	65
$\mathbf{S}_\alpha(F)$	122
$\mathbf{S}_\alpha^n(F)$	123

Index

	Page
affine transformation	81
axiom of minimal element	67
basis	29
bounded figure	66
characteristic equation	31
characteristic value	30
characteristic vector	30
convex figure	70
convex hull	89
Five-Point Problem	55
Grand Problem I	16
Grand Problem II	16
homothety	79
integral dependence	8
integral independence	8
Intermediate Value Theorem	71
isometry	79
linear combination	29
linear dependence	28
linear independence	27
One-Hundred-Dollar Problem	96

parallel projection   81
pigeonhole principle   52
span (vector space)   29
supporting line   74
trigon   152
tool   20
vector   27
vector space $\mathbf{R}^3$ 27